Crop Residue Management: For Soil Health, Crop Productivity & Environment Quality

Crop Residue Management: For Soil Health, Crop Productivity & Environment Quality

Editor

Rajani Satam

Crop Residue Management: For Soil Health, Crop Productivity & Environment Quality

Edited by **Rajani Satam**

Printed in 2017

ISBN: 978-1-68117-056-5

Library of Congress Control Number: 2015935454

Contents

Preface

The management of croplands has a large impact on the quantity and quality of food and fiber production and on air and water quality, all of which influence the quality of our environment. Management of the non-harvested plant tissues, such as leaves, stems, branches, and roots that constitute the residues from the production of food and fiber, is one of the farm practices that affects crop production via nutrient availability and cycling. Management of these residues also provides opportunities for control of dust in the air and sediments and nutrients in waters through control of soil erosion caused by wind and water. Thus, the task ahead for the management of croplands for improvements in the overall well-being of people involves the application of known effective crop residue management systems to maintain crop production and to reduce wind and water erosion. There is need, as well, for development of more effective management systems through laboratory and field research. Soil is the basis of farming. It delivers water and nutrients to crops, physically supports plants, helps control pests, determines where rainfall goes after it hits the earth, and protects the quality of drinking water, air, and wildlife habitat. The goal of soil management is to protect soil and enhance its performance, so you can farm profitably and preserve environmental quality for decades to come. One cannot control slope, texture, climate, and other critical soil factors. But one can control tillage, crop rotations, soil amendments, and other management choices. Through these choices you change the structure, biological activity, and chemical content of soil, and you influence erosion rates, pest populations, nutrient availability, and crop production.

Editor

Chapter 1

A Review on Recycling of Sunflower Residue for Sustaining Soil Health

Subhash Babu[1], D.S. Rana[2], G.S. Yadav[3],
Raghavendra Singh[1], and S.K. Yadav[4]

[1]ICAR Research Complex for NEHR, Sikkim Centre, Tadong, Gangtok,
East Sikkim 737102, India

[2]Indian Agricultural Research Institute, New Delhi 110012, India

[3]ICAR Research Complex for NEHR, Tripura Centre, Lembucherra,
West Tripura 799210, India

[4]Central Potato Research Station, Shillong 793009, India

ABSTRACT

Modern agriculture is now at the crossroads ecologically, economically,
technologically, and socially due to soil degradation. Critical analysis
of available information shows that problems of degradation of soil

health are caused due to imbalanced, inadequate and promacronutrient fertilizer use, inadequate use or no use of organic manures and crop residues, and less use of good quality biofertilizers. Although sizeable amount of crop residues and manure is produced in farms, it is becoming increasingly complex to recycle nutrients, even within agricultural systems. Therefore, there is a need to use all available sources of nutrients to maintain the productivity and fertility at a required level. Among the available organic sources of plant nutrients, crop residue is one of the most important sources for supplying nutrients to the crop and for improving soil health. Sunflower is a nontraditional oil seed crop produced in huge amount of crop residue. This much amount of crop residues is neither used as feed for livestock nor suitable for fuel due to low energy value per unit mass. However, its residue contains major plant nutrients in the range from 0.45 to 0.60% N, 0.15 to 0.22% P, and 1.80 to 1.94% K along with secondary and micronutrients, so recycling of its residue in the soil may be one of the best alternative practices for replenishing the depleted soil fertility and improving the physical, chemical, and biological properties of the soil in the present era of production. However, some researchers have reported allelopathic effects of sunflower residue on different crops. So, selection of suitable crops and management practices may play an important role to manage the sunflower residue at field level.

INTRODUCTION

Nowadays, modern agriculture is facing the problem of land and water degradation, environmental pollution, lowering of water table, and global competition. There is a decrease in factor productivity, development of multinutrient deficiency, build of obnoxious weeds and pests, and increasing cost of production over the last two decades. This means that modern agriculture is now at the crossroads ecologically, economically, technologically, and socially; our present growth rate in agriculture is not keeping pace with the population growth rate. Our greatest living industry is in distress. Increasing the productivity and profitability of small farmers in an economically sustainable manner is the most effective step for reducing poverty and hunger in our country. The major pathway has to be productivity enhancement. Among the various reasons for this alarming situation, physical, chemical, and biological deterioration of soil ranks at the top. Critical analysis of available information shows that problems of degradation of soil health are caused due to imbalanced, inadequate and promacronutrient fertilizer use, inadequate use or no use of organic manures and crop residues, and less use of good quality bio-fertilizers. With intensive cropping, nutrient removal by crops from soil (34 mt) has far exceeded replenishment (18 mt). Present estimates show a deficit of 16 mt of plant nutrients, which is likely to grow further with intensification in agriculture and increasing soil degradation. The fertilizer consumption ratio of 8.9 : 2.8 : 1.0 as against 4 : 2 : 1 indicates an erratic and imbalanced fertilizer use, causing decline in soil fertility and productivity. The imbalance use of fertilizer has also been responsible for low fertilizer response. Even if, so-called balanced fertilization emphasis on NPK will not be able to sustain high productivity. Multi-nutrient deficiency including micronutrients is becoming very common. Therefore, there is a need to use all available sources of nutrients to maintain the productivity and fertility at a required level. The degradation of soil fertility owing to over mining of nutrients and inadequate replenishment through fertilizers and other sources can only be curbed through adoption of INM technology. Among the available organic sources of plant nutrients, crop residue is one of the most important sources for supplying nutrients to the crop and for improving soil health. Integrated nutrient management based on low cost and locally available organic sources is more rational, sustainable, and economical. Amongst the nine oilseed crops, sunflower (Helianthus annuus L.) has the potential and can play an important role in meeting out shortage of oilseeds in the country. Sunflower (Helianthus annuus L.) is a nontraditional crop introduced in India during the year 1969. It now occupies an important place in all agroclimatic zones. It holds a great promise because of its short duration, photo insensitivity, and wide adaptability to different agroclimatic regions and soil types. It can be grown at any time of the year and can serve as an ideal crop during the period when the land is otherwise fallow. From the 1500 ha in 1972-73, its area increased to 2.35 mha in 2007-08. Spring sunflower in the north-west part of India has potential to yield 4–6 t/ha crop residue and 2–2.5 t/ha seed yield and it has less water, nutrient, and other input requirement than wheat. This much amount of crop residues is neither used as feed for livestock nor suitable for fuel due to low energy value per unit mass. However, its residue contains major plant nutrients in the range from 0.45 to

0.60% N, 0.15 to 0.22% P, and 1.80 to 1.94% K along with secondary and micronutrients, so recycling of its residue in the soil may be one of the best alternative practices for replenishing the depleted soil fertility and improving the physical, chemical, and biological properties of the soil. Some researchers have reported allelopathic effects of sunflower residue on different crops, which put a question on choice of crop after sunflower and its residue incorporation [1–3]. Schon and Einhelling [4] demonstrated that incorporation of dried sunflower leaf material into the soil inhibited germination and growth of grain sorghum. Water soluble toxic substances could leak from the plant and from decomposing residue causing allelopathic interference. Leachates from the plants have been shown to suppress seed germination and vegetative propagules and early seedling growth [5, 6] and decrease radicle growth [7]. Aqueous extract of some plants inhibits seedling growth [8], root and shoot growth [9], and germination [10] and induces mortality of plants [11]. The effects may be due to a variety of processes that may include reduced cell division in the roots, suppressed hormonal activity, reduced ion uptake, inhibition of protein synthesis, inhibition of photosynthesis and respiration, inhibition of enzyme activity, and reduced cell membrane permeability [12]. During the decomposition of crop residue, several acids and other compounds are released, which affect the soil pH, nutrients availability, microbial population, and enzymatic activities, namely, dehydrogenase and phosphatase in soil.

EFFECT OF SUNFLOWER RESIDUE RECYCLING ON GROWTH AND YIELD OF CROPS

Inhibitory or stimulatory effects of crop residue on germination and establishments of crops caused by residues of either crops or weeds have led to investigation of the release of toxic or growth stimulating compounds from such residues. In this regards, residues of sunflower have been examined for their potential to reduce or stimulate the growth of succeeding crops in a system. Isidron et al. [13] reported that allelochemicals of mature maize foliage had an inhibitory effect on the sunflower, maize, and sorghum. Srisa-Ard [14] also indicated that crop residues derived from roots of both sunflower and soybean plants had significant inhibitory effects on plant height, root dry weight, top growth dry weight, and total dry weight of the sunflower plants. In a bioassay study, Ashrafi et al. [15] reported that sunflower extracts reduced wild barley hypocotyl length, hypocotyl weight, radicle weight, seed germination, and radicle length by as much as 44, 58, 61, 69, and 79%, respectively, when compared with a water control. Increasing the water extract concentrations from 4 to 20/100 mL water of all sunflower parts significantly increased the inhibition of wild barley germination, seedling length, and weight. In Uttaranchal, Pal and Sand [16] reported that the precedingspring sunflower reduced the plant population, dry matter production, and all yield attributes of succeeding crops, that is, rice, maize, soybean, and pigeonpea. In greenhouse studies at Hissar, Narwal et al. [17] reported that among the different parts of sunflower (stem, leaf, inflorescence, and root), stems showed the highest inhibitory effect on the growth and yield of wheat, followed by the roots, leaves, and inflorescences. Increasing doses of sunflower biomass resulted in significant reductions in the seedling growth and yield of wheat. Chlorogenic acid, caffeoylquinic acid, and neochlorogenic acid in sunflower stems at varying concentrations might be responsible for reduced growth and yield in wheat. Similar results with different parts of sunflower residue on different crops in field studies were also reported by several workers ([12] on Bambara groundnut, [18, 19] on lobia, sorghum, bajra, and maize, [20] on mung bean and pearl millet, [21] on wheat, and [1] on black gram, green gram, red gram, finger millet, pearl millet, maize, sorghum, gingelly, and groundnut).

A study conducted by Gill and Sandhu [22] indicated that ground sunflower leaves from a matured crop incorporated into the soil (0.5–2.5% weight/weight basis) decreased sunflower seed germination at all concentrations. However, only concentrations >2% w/w decreased seed germination of the maize, cotton, pigeonpea, soybean, and pearl millet. While the shoot and root growth responses to allelopathic effects were dependent on the species, the adverse effects on growth of all the species were evident at the higher concentrations. Narwal et al. [23] reported that incorporation of wheat straw reduced the growth and yield of succeeding sorghum, pearl millet, maize, cluster bean, and cowpea. In pot studies, Narwal et al. [2, 3] reported that sunflower-infested soil, that is, from sunflower rhizospheres, inhibits the germination, growth, development, and grain yields and induced seedling mortality of all tested crops (sorghum, pearl millet, maize, cotton, cowpea, cluster bean, green gram, black gram, soybean, and pigeonpea).

At IARI New Delhi, Rana et al. [24] reported that sunflower as preceding crop caused 51 per cent reduction in the seed yield of chickpea compared to maize as a preceding crop. The yield-reducing effect of sunflower was not observed on wheat and Indian mustard. The adverse effect of complete sunflower residue incorporation on the productivity of succeeding crops was more compared to partial or complete residue removal, but there was gradual improvement in the yield of these crops under complete residue over the years. In the last harvest wheat and Indian mustard gave 3.61 and 1.88 t/ha seed yield under complete residue incorporation which was 13.5 and 13.2 per cent higher than seed yield recorded under complete residue removal. Kaya et al. [25] investigated the effect of biodegradation products of sunflower heads (BPSH) at various concentrations (1.0%, 2.5%, 10%, and 100%) on the seed germination and some growth parameters of rajmash, chickpea, and wheat and reported that the percentage of seed germination and germination index of seeds were similar between the control and 1.0% and 2.5% BPSH groups, but these values decreased at higher concentrations. On the other hand, growth of the seedlings gradually increased up to a concentration of 10% BPSH and decreased at 100% concentration. As a result, at concentrations up to 10% the product was found to be beneficial for growth of plants. Mangare et al. [26] stated that incorporation of sunflower straw along with inorganic fertilizer increased the grain yield of green gram and sunflower as well as straw yield over control. The pooled analysis indicated that the highest grain yield was observed with the application of sunflower straw @ 4 t/ha + 125% N + 100% P of RDF, followed by the application of sunflower straw @ 4 t/ha + 100% of RDF which were found at par with former. The highest benefit cost ratio was obtained with the application of crop residues treatments as compared to control treatment. Badnur et al. [27] also found that sunflower stalks at 5 t/ha + subabul loppings (50:50) gave the highest grain yield (1408 kg/ha), followed by recommended rate of fertilizer (1403 kg/ha) and sunflower threshed ear heads at 5 t/ha + 20 kg N/ha (1399 kg/ha). The lowest grain yield (772 kg/ha) was recorded from sunflower stalks at 5 t/ha. Treatment with sunflower threshed ear heads at 5 t/ha and recommended rate of fertilizer recorded the second highest sorghum fodder yield, 3472 and 3135 kg/ha, respectively. Sunflower threshed ear heads at 2.5 t/ha + subabul loppings at 2.5 t/ha (50:50), sunflower stalks at 5 t/ha + subabul loppings (50:50), and sunflower threshed ear heads at 5 t/ha + 20 kg N/ha resulted in the third highest 1,000-grain weights, 35.86, 35.47 and 35.33 g, respectively.

EFFECT OF SUNFLOWER RESIDUE RECYCLING ON SOIL BIOLOGICAL PROPERTIES

One of the most important challenges facing humanity today is to conserve/sustain natural resources, including soil and water, for increasing food production while protecting the environment. As the world population grows, stress on natural resources increases, making it difficult to maintain food security. Long-term food security requires a balance between increasing crop production, maintaining soil health and environmental sustainability. Sustainability of agricultural systems has become an important issue all over the world. Many issues of sustainability are related to soil quality and its change with time [28]. According to Doran and Parkin [29], soil quality is "the capacity of a soil to function within ecosystem boundaries to sustain biological productivity, maintain environmental quality, and promote plant and animal health." Soil biological activities have been suggested as one of the important indicators of soil quality [30]. Decomposition of plant residues is the microbially mediated progressive breakdown of organic material into C (biomass or CO_2) and other nutrients [31]. Crop residues decompose into two distinct phases, an initial rapid phase, in which about 70% of C initially present in the residues is lost as CO_2, followed by a slower phase during which the more resistant fraction is decomposed [32]. Immature plant residues with a high concentration of water-soluble compounds such as sugars, amino acids, and organic acids are decomposed more rapidly than mature material, which contain a higher proportion of resistant compounds such as cellulose, lignin, phenols, or waxes. Residue factors include chemical composition, C/N ratio, lignin content, and the size of residue particles [33]. Residue C/N ratio is a common indicator of residue quality but is not necessarily an accurate predictor of decomposition rate [34]. The incorporation of crop residues into the soil modifies its chemical and biochemical properties, including soil-enzyme activity [35–37], the behaviour of which has often been related to the amount [38–40] as well as to the type of organic matter [41, 42]. Soil enzymes play a major role in nutrient availability [43]. In soils, enzymes may be associated with viable cells, dead cells (abiontic enzymes), and cell debris and immobilized enzymes in the soil matrix [44]. Since, soil-enzyme systems are associated with organic residue management, the burying of crop residues into the soil not only plays an important role in the soil's chemical and biochemical environment, but also affects the rate at which nutrients become available to crop plants as well as to other forms of life in the soil. Therefore, any management practice that influences the biological populations of soil would be expected to produce changes in soil enzyme activity levels. The effect exerted on soil-dehydrogenase and soil phosphatase depends on the type of crop residues incorporated. Dehydrogenase is considered to play an important role in the initial stages of the oxidation of soil organic matter [45] by transferring hydrogen and electrons from substrates to acceptors. Soil water content and temperature influence the dehydrogenase activity indirectly by affecting the soil oxidation-reduction status [46]. Addition of tobacco and sunflower residues in soil increases the activities of most of the soil enzymes, while tomato residues increased only the amylase and phosphodiesterase activities. As the enzyme activities were positively correlated to each other, a common source of the enzymes is suggested even though the coefficients of correlation demonstrate that only a low percentage of the variability can be ascribed to the interactions among enzyme activities [42]. Soil enzymes are used as biological indices of soil fertility under different tillage practices. Phosphatases are inducible enzymes excreted by plant roots and soil organisms, which can be stimulated by P starvation. Therefore, phosphatase activity has been regarded as an important factor in maintaining and controlling mineralization rate of soil organic P and a good indicator of P deficiency [47]. The measurements, of phosphatase enzyme activity performed during three years of study, were significantly lower in the soil under winter wheat grown in the conventional-monoculture farming system, in comparison to organic system. The significantly lower phosphatase enzyme activity in monoculture soil was related to lower microbial biomass carbon in the soil compared to organic system [48]. At Hisar, Kaur and Kapoor [49] reported that addition of N to residue to have a C:N ratio of 40:1 accelerated residue decomposition. An amount of 61 and 58.7% of incorporated sunflower residues was decomposed in 12 weeks when residues (C:N ratio 73.1) were added at 1.5 and 3.0t/ha, respectively. The extent of decomposition increased to 66.6 and 61.1% when residues were incorporated at 1.5 and 3.0t/ha, respectively, along with N application to adjust C:N ratio of residues to 40:1. Microbial biomass C increased with residues as well as N application up to 3 weeks and then it declined. Dehydrogenase activity during decomposition of sunflower residues was higher in first week after that it declined. At Akola, Maharashtra, Ravankar et al. [50] reported that incubation of soil with 1% organic residues of cotton stalk, safflower straw, sorghum stubble, soybean stover, wheat straw, sugarcane trash, groundnut husk, sunflower straw, green gram stover, Parthenium with seed, grass complex with seed, and Xanthium with seed mixed showed wide variation in the rate of decomposition, C:N ratio and microbial population at different intervals. Carbon dioxide evolution was maximum during the first 15 days and it decreased thereafter. The rate of decomposition was the highest for groundnut husk as compared to other residues. After incubation for 30 days, the lowest C:N ratio was observed in grass complex. Fungal, bacterial, and actinomycetes populations increased at 30 days of incubation. Bacteria were predominant over fungi and actinomycetes.

EFFECT OF SUNFLOWER RESIDUE RECYCLING ON SOIL CHEMICAL PROPERTIES

A good account of work has been done on the use of crop residue to supplement the nutrient requirement of cropping system and to improve soil health. Sharma et al. [51] stated that nutrient-rich residues of the castor and sunflower are mostly burnt because of their high C/N ratio. These high C/N ratio residues can be recycled successfully, if they are supplemented with other low C/N-ratio farm-based organics and some chemical additives. Application of crop residues with a high C:N ratio often leads to adverse impacts on available N in soil and growth of crops planted immediately after crop residue incorporation. A large number of organic compounds, particularly phenolic acid and acetic acid, are released during the decomposition of crop residues. The application of crop residues can cause short-term immobilization of both P and S, particularly in aerobic soils. Crop residues contain large amounts of K, which upon incorporation increased K availability in soil and helped to reduce K depletion from nonexchangeable K fraction of soil. Long-term application of crop residues increased the organic matter, total N content, and availability of several nutrients in soils. The rate of increase in soil organic matter is low due to high turnover rates of C under tropical conditions. The increase in soil organic matter levels due to crop residue recycling was determined by the duration, amount, and quality of residue, soil type, climatic conditions, and cropping system followed. Crop residues influence the chemical and biological properties of the soil [52].

Ordonez-Fernandez et al. [53] conducted a field experiment in southern Spain and reported that the pea residue supplied the highest amount of nitrogen to the soil throughout its decomposition cycle; it lost 76.6% of its initial content in nitrogen, compared to the 48 and 56% of N released by wheat residues and sunflower, respectively. At the beginning of its decomposition cycle, the wheat residue had the lowest mass and gave the most cover, with values of 65%, which was 8.6% and 20.2% more than the cover estimated for the pea and sunflower residues, respectively. The sunflower residue lasted longest, only losing 18% of its initial cover over 109 days of decomposition, compared to 47% of wheat and 53% for pea. The amount of carbon released was similar for the three residues and was around 500 kg/ha. Corbeels et al. [54] studied the C mineralization and N transformations during the decomposition of sunflower stalks and wheat straw with and without addition of $(NH_4)_2SO_4$ in a Vertisol. Soil samples were incubated under aerobic conditions for 224 days at 22°C. The plant residues were added at a rate of 5.2 g/kg soil. Nitrogen was applied at a rate of 50.7 mg N/kg soil. Gross N immobilization and mineralization were calculated on the basis of the isotopic dilution technique. At the end of the incubation period a ^{15}N balance was established. Respectively, 68 and 45% of the applied residue-C mineralized from the sunflower stalks and wheat straw after 224 days. Both crop residues caused losses of up to 25% of added ^{15}N after 224 days of incubation. These ^{15}N losses were about three times larger than in the control soil and were probably due to denitrification. The net immobilization of soil derived N following residue incorporation was the largest in the case of wheat straw, depleting all soil inorganic N. In the wheat straw treatment with added $(NH_4)_2SO_4$ soil inorganic N remained available, resulting in an enhanced initial C mineralization and N immobilization compared to the treatment without added N. In the case of the sunflower stalks, the high inorganic N content of the stalks suppressed the effects of N addition on C mineralization and N immobilization/mineralization. Gross N immobilization amounted to 31.9 and 28.2 mg N/g added C after 14 days for wheat straw and sunflower stalks, respectively. At the end of the incubation, about 35% of the newly immobilized N was re-mineralized in both plant residue treatments. Gross N immobilization plotted against decomposed C suggests that fairly uniform C-N relationships exist during the decomposition of divers C substrates. Immobilization of available N and P during initial phase of decomposition of sunflower residue was also reported by Kaur and Kapoor [49]. Decreases in Olsen-P in soil due to incorporation of organic manure and crop residues having wider C:N ratio were also reported by Jalali [55]. In laboratory conditions, Das et al. [56] reported that incorporation of loppings of perennial pigeonpea, sorghum stover, and FYM caused initial immobilization of P. However, no P immobilization was observed with mung bean, Leucaenaresidues, and groundnut shells at 0.5 FC. At FC, none of the residues caused immobilization. The quantity of P released was significantly related to the amount of P added through residues. In contrast, Das and Puste [57] showed that the amounts of ammoniacal-N (NH_4-N), nitrate-N (NO_3-N), hydrolysable N (HL-N), and nonhydrolysable (NHL-N) were increased for up to 60 days of soil submergence and increased further with the increase (1% by weight of soil) of various organic waste materials like crop residues, well-decomposed cow dung, composts, and other rural and urban wastes application. Similarly, Raut et al. [58] in his study showed that incorporation of sunflower straw @ 4 t/ha + RDF @ (125% N + 100% P) in green gram recorded significantly higher soil nitrogen, phosphorus, and potassium content in green gram-sunflower sequence. Content of micronutrient (Zn, Fe, Cu, Mn, B, and Mo) including sulphur was also maximum with the incorporation of sunflower straw @ 4 t/ha + RDF @ (125% N + 100% P) in green gram. Similar results were obtained by Santamaría et al. [59] with the incorporation of sunflower hulls, a residue of oil industries, in the upper layer of soil as organic amendment. In contrast, Jalali and Ranjbar [60] observed that the rates of P release of the residues (sunflower and wheat) were considerably higher during the first 4 weeks of incubation than during the second phase of incubation (weeks 5–12). Phosphorus release by residues was similar to the decomposition pattern. They also opined that residue P content was correlated with P release, but not with decomposition rate. Gong et al. [61] reported that sunflower oil residue incorporation in soil decreases the soil pH. In another study Badnur et al. [27] reported that incorporation of sunflower residues either with subabul loppings or inorganic N recorded higher available nutrients than incorporation of sunflower residues alone. However, available N and K were highest with sunflower threshed ear heads at 2.5 t/ha + subabul loppings @ 2.5 t/ha (50:50), (222 and 450 kg/ha, resp.), while available P was highest upon treatment with sunflower stalks at 5 t/ha + subabul loppings (50:50), (24.7 kg/ha). In maize-wheat cropping system at Kanpur, Prasad et al. [62] found that application of FYM and gypsum reduced soil pH by 0.38 and 0.30 units, EC by 0.07 and 0.06 units than their initial values of 7.8 and 0.37, respectively. Organic carbon increased by 22.6 and 15.1%, available N increased by 41.2 and 29.1%, available P increased by 53.8 and 33.3%, and available K increased by 11.7 and 6.1% under treatments of FYM and gypsum, respectively, than their initial values. In the case of fertilizer, 100, 75, and 50% of recommended doses showed reduction in soil pH by 0.30, 0.30, and 0.28 units and in EC by 0.07, 0.06, and 0.05 units, respectively, than their initial pH value of 7.8 and initial EC of 0.37. Application of recommended dose, 75% of RDF and 50% of RDF increased other parameters such as organic carbon by 18.9, 15.1, and 13.2%, available N by 44.7, 31.9, and 22.0%, available P by 49.7, 34.9, and 20.5%, and available K by 11.7, 6.9, and 2.6%, respectively. The nutrients release from decomposing residues of rape, sunflower, and soybean in amended soil under laboratory condition was investigated by Scagnozzi et al. [63] at Pisa, Italy, and reported that total N, available P, exchangeable K^+, Ca^{2+}, and Mg^{2+} in all the amended samples increased significantly. Generally, the increase in the amounts of these nutrients was maintained until the end of the incubation period; the mineralization of the three crop residues enhanced soil fertility. In amended soil samples, NH_4-N disappeared within 14 days, while available N was released as NO_3-N after 60 days in soybean-treated and after 120 days in rape- and sunflower-treated soil, respectively.

Batish et al. [18, 19] conducted an experiment to investigate the effect of residues of the noxious weedParthenium hysterophorus in soil as well as under laboratory conditions. Soils were infested with different amounts of Parthenium hysterophorus residues to determine the changes in soil chemistry and revealed that the pH of all the modified soils decreased, whereas the conductivity, organic carbon, and organic matter increased. The amount of sodium and potassium increased, whereas that of zinc decreased. In the soil infested with 4 g of Parthenium hysterophorus residue, the amount of available nitrogen decreased. Paul et al. [64] summarized that the addition of plant residue in soil results in a rapid (days 0–7) increase of soil pH due to the association, and particularly oxidation, of added organic anions. This was followed by a gradual (days 7–119) pH decline attributed to the mineralization and subsequent nitrification of added organic N. The addition of 12.5–25.0 g of cereal crop residues/kg of soil and 6.25–25.0 g of legume-based pasture residues/kg of soil resulted in a net alkalization of the surface 2.5 cm of soil. The magnitude of such gradients will be particularly large with the return of large quantities of plant residues of high ash alkalinity in soils of relatively low initial pH and biological activity, and when the surface of the soil is exposed to moist-dry cycles.

CONCLUSIONS

A review of the literature clearly indicates that sunflower residue incorporation had the adverse effect on crops due its allelocompounds and reduced the growth and yield. Sunflower residue is usually considered a problem but when managed correctly it can improve soil organic matter dynamics and nutrient cycling, thereby creating a rather favorable environment for plant growth on long-term basis. Sunflower residue contains large quantities of nutrients, and thus the return of sunflower stover to the soil can save a considerable quantity of fertilizers. So, on the basis of the above review it can be concluded that recycling of sunflower residue improves the soil biological, chemical, and physical properties, which may enhance the agricultural sustainability in the near future.

REFERENCES

1. G. Velu, "Crop response to allelopathic effect of sunflower," Research and Development Reporter, vol. 6, no. 1, pp. 16–21, 1989.

2. S. S. Narwal, T. Singh, J. S. Hooda, and M. K. Kathuria, "Allelopathic effects of sunflower on succeeding summer crops. I. Field studies and bioassays," Allelopathy Journal, vol. 6, no. 1, pp. 35–48, 1999.

3. S. S. Narwal, S. Yadava, and S. Gupta, "Allelopathic effects of sunflower on succeeding summer crops. 2. Pot culture and biomass decomposition," Allelopathy Journal, vol. 6, no. 2, pp. 209–226, 1999.

4. M. K. Schon and F. A. Einhelling, "Allelopathic effects of cultivated sunflower on grain sorghum,"Botanical Gazette, vol. 143, pp. 505–510, 1980.

5. R. C. Babu and O. S. Kandasamy, "Allelopathic effect of Eucalyptus globulus Labill. on Cyperus rotundusL. and Cynodon dactylon L. Pers," Journal of Agronomy and Crop Science, vol. 179, no. 2, pp. 123–126, 1997.

6. S. R. Dhawan and S. K. Gupta, "Allelopathic potential of various leachate combinations towards SG and ESG of Parthenium hysterophorous Linn," World-Weeds, vol. 3, no. 1, pp. 85–88, 1996.

7. C. M. Casado, "Allelopathic effect of Lantana camara (Verbanaceae) on morning glory (Ipomoea tricolor)," Rhodora, vol. 97, no. 891, pp. 264–274, 1995.

8. J. Lydon, J. R. Teasdale, and P. K. Chen, "Allelopathic activity of annual wormwood (Artemisia annua) and the role of artemisinin," Weed Science, vol. 45, no. 6, pp. 807–811, 1997.

9. D. P. Athanassova, "Allelopathic effect of Amaranthus retroflexus L. on weeds and crops," in Seizième conférence du COLUMA. Journées internationales sur la lutte contre les mauvaises herbes, pp. 437–442, Reims, France, 1996.

10. J. E. Pratley, P. Dowling, and R. Medd, "Allelopathy in annual grasses," in Wild Oats, Annual Ryegrass and Vulpia: Proceeding of a Workshop Held at Orange, vol. 1, pp. 213–214, New South Wales, Australia, 1996.

11. M. Eyini, A. U. Maheswari, T. Chandra, and M. Jaykumar, "Allelopathic effects of leguminous plants leaf extracts on some weeds and corn," Allelopathy Journal, vol. 3, no. 1, pp. 85–88, 1996.

12. U. Batlang and D. D. Shushu, "Allelopathic activity of sunflower (Helianthus annuus L.) on growth and nodulation of bambara groundnut (Vigna subterranea (L.) Verdc.)," Journal of Agronomy, vol. 6, no. 4, pp. 541–547, 2007.

13. M. P. Isidron, C. D. Iglesias, and H. G. Rodriguez, "Study of the allelopathic effects of residues of maize and sorghum at different physiological stages on crops of economic importance and record of the entomofauna present in sunflower (Helianthus annuus L.)," Centro Agricola, vol. 27, no. 4, pp. 29–32, 2000.

14. K. Srisa-Ard, "Effects of crop residues of sunflower (Helianthus annuus), maize (Zea mays L.) and soybean (Glycine max) on growth and seed yields of sunflower," Pakistan Journal of Biological Sciences, vol. 10, no. 8, pp. 1282–1287, 2007.

15. Z. Y. Ashrafi, S. Sadeghi, and H. R. Mashhadi, "Allelopathic effects of sunflower (Helianthus annuus) on germination and growth of wild barley (Hordeum spontaneum)," in Proceedings of the 9th International Conference on Precision Agriculture, pp. 324–336, Denver, Colo, USA, July 2008.

16. M. S. Pal and N. K. Sand, "Effects of sunflower (Helianthus annuus L.) on growth and yield of succeeding crops," Allelopathy Journal, vol. 17, no. 2, pp. 297–302, 2006.

17. S. S. Narwal, R. Palaniraj, S. C. Sati, and L. S. Rawat, "Effects of different parts of sunflower (Helianthus annuus) biomass on wheat (Triticum aestivum)," Journal of Ecobiology, vol. 15, no. 5, pp. 371–376, 2003.

18. D. R. Batish, H. P. Singh, J. K. Pandher, V. Arora, and R. K. Kohli, "Phytotoxic effect of Parthenium residues on the selected soil properties and growth of chickpea and radish," Weed Biology and Management, vol. 2, no. 2, pp. 73–78, 2002.

19. D. R. Batish, P. Tung, H. P. Singh, and R. K. Kohli, "Phytotoxicity of sunflower residues against some summer season crops," Journal of Agronomy and Crop Science, vol. 188, no. 1, pp. 19–24, 2002.

20. K. Kaur and K. K. Kapoor, "Effect of incorporation of sunflower residues in soil on germination of mungbean and pearl millet," Environment and Ecology, vol. 17, no. 3, pp. 693–695, 1999.

21. P. J. Morris and D. J. Parrish, "Effects of sunflower residues and tillage on winter wheat," Field Crops Research, vol. 29, no. 4, pp. 317–327, 1992.

22. D. S. Gill and K. S. Sandhu, "Studies on allelopathic effect of sunflower (Helianthus annuus) residues on succeeding kharif crops," Indian Journal of Ecology, vol. 20, no. 2, pp. 169–172, 1993.

23. S. S. Narwal, M. K. Sarmah, and D. P. S. Nandal, "Allelopathic effects of wheat residues on growth and yield of fodder crops," Allelopathy Journal, vol. 4, no. 1, pp. 111–120, 1997.

24. D. S. Rana, G. Giri, K. S. Rana, and D. K. Pachauri, "Effect of sunflower (Helianthus annuus) residue management on productivity, economics and nutrient balance sheet of sunflower- and maize (Zea mays)-based cropping systems," Indian Journal of Agricultural Sciences, vol. 74, no. 6, pp. 305–310, 2004.

25. Y. Kaya, M. Şengül, H. Öütçü, and O. F. Algur, "The possibility of useful usage of biodegradation products of sunflower plants," Bioresource Technology, vol. 97, no. 4, pp. 599–604, 2006.

26. P. N. Mangare, S. M. Shendurse, S. G. Matale, and S. P. Nandapure, "Effect of incorporation of crop residue on crop productivity and economics under green gram-sunflower sequence," Green Farming, vol. 2, no. 1, pp. 47–49, 2008.

27. V. P. Badnur, M. A. Bellakki, and S. I. Tolnur, "Incorporation of sunflower crop residues for integrated nutrient management of rabi sorghum in Vertisol," Karnataka Journal of Agricultural Sciences, vol. 13, no. 3, pp. 733–734, 2000.

28. D. L. Karlen, M. J. Mausbach, J. W. Doran, R. G. Cline, R. F. Harris, and G. E. Schuman, "Soil quality: a concept, definition, and framework for evaluation," Soil Science Society of America Journal, vol. 61, no. 1, pp. 4–10, 1997.

29. J. W. Doran and T. B. Parkin, "Defining and assessing soil quality," in Defining Soil Quality for a Sustainable Environment, J. W. Doran, D. C. Coleman, D. F. Bezdicek, and B. A. Stewart, Eds., Soil Science Society of America Special Publication no. 35, pp. 3–21, Soil Science Society of America, Madison, Wis, USA, 1994.

30. R. P. Dick, "Soil enzymatic activities as in indicator of soil quality," in Definding Soil Quality for a Sustainable Development, J. W. Doran, D. C. Coleman, D. F. Bezdicek, and B. A. Stewart, Eds., pp. 107–124, Soil Science Society of America, Madison, Wis, USA, 1994.

31. K. Kumar and K. M. Goh, "Crop residues and management practices: effects on soil quality, soil nitrogen dynamics, crop yield, and nitrogen recovery," Advances in Agronomy, vol. 68, pp. 197–319, 1999.

32. W. J. Wang, J. A. Baldock, R. C. Dalal, and P. W. Moody, "Decomposition dynamics of plant materials in relation to nitrogen availability and biochemistry determined by NMR and wet-chemical analysis,"Soil Biology and Biochemistry, vol. 36, no. 12, pp. 2045–2058, 2004.

33. J. M.-. Johnson, N. W. Barbour, and S. L. Weyers, "Chemical composition of crop biomass impacts its decomposition," Soil Science Society of America Journal, vol. 71, no. 1, pp. 155–162, 2007.

34. E. Handayanto, G. Cadisch, and K. E. Giller, "Nitrogen release from prunings of legume hedgerow trees in relation to quality of the prunings and incubation method," Plant and Soil, vol. 160, no. 2, pp. 237–248, 1994.

35. J. W. Doran, "Microbial changes associated with residue management with reduce management with reduced tillage," Soil Science Society of America Journal, vol. 44, no. 3, pp. 518–524, 1980.

36. J. M. Duxbury and R. L. Tate III, "The effect of soil depth and crop cover on enzymatic activities in Pahokee Muck," Soil Science Society of America Journal, vol. 45, no. 2, pp. 322–328, 1981.

37. W. A. Dick, N. G. Juma, and M. A. Tabatabai, "Effects of soils on acid phosphatase and inorganic pyrophosphatase of corn roots (Zea mays)," Soil Science, vol. 136, no. 1, pp. 19–25, 1983.

38. T. W. Speir and D. J. Ross, "Biochemical changes with pasture development in a West Coast wet land soil: a note," New Zealand Journal of Science, vol. 26, no. 4, pp. 505–508, 1983.

39. J. F. Power and J. O. Legg, "Effect of crop residues on the soil chemical environment and nutrient availability," in Crop Residue

Management Systems, W. R. Oschwald, Ed., ASA Special Publication, pp. 80–110, American Society of Agronomy, 1978.

40. T. M. Klein and J. S. Koths, "Urease, protease and acid phosphatase in soil continuously cropped to corn by conventional or no-tillage method," Soil Biology and Biochemistry, vol. 12, no. 3, pp. 293–294, 1980.

41. R. C. Dalal, "Effect of plant growth and addition of plant residues on the phosphatase activity in soil,"Plant and Soil, vol. 66, no. 2, pp. 265–269, 1982.

42. P. Perucci, L. Scarponi, and M. Businelli, "Enzyme activities in a clay-loam soil amended with various crop residues," Plant and Soil, vol. 81, no. 3, pp. 345–351, 1984.

43. D. A. Martens, J. B. Johanson, and W. T. Frankenberger, "Production and persistence of soil enzymes with repeated addition of organic residues," Soil Science, vol. 153, no. 1, pp. 53–61, 1992.

44. R. G. Burns, "Enzyme activity in soil: location and a possible role in microbial ecology," Soil Biology and Biochemistry, vol. 14, no. 5, pp. 423–427, 1982.

45. D. J. Ross, "Some factors influencing the estimation of dehydrogenase activities of some soils under pasture," Soil Biology and Biochemistry, vol. 3, no. 2, pp. 97–110, 1971.

46. M. Brzezińska, Z. St.pniewska, and W. St.pniewski, "Soil oxygen status and dehydrogenase activity,"Soil Biology and Biochemistry, vol. 30, no. 13, pp. 1783–1790, 1998.

47. C. P. Vance, C. Uhde-Stone, and D. L. Allan, "Phosphorus acquisition and use: critical adaptations by plants for securing a nonrenewable resource," New Phytologist, vol. 157, no. 3, pp. 423–447, 2003.

48. A. Gajda and S. Martyniuk, "Microbial biomass C and N and activity of enzymes in soil under winter wheat grown in different crop management systems," Polish Journal of Environmental Studies, vol. 14, no. 2, pp. 159–163, 2005.

49. K. Kaur and K. K. Kapoor, "Carbon mineralization, microbial biomass and nutrient release pattern during decomposition of sunflower residues in soil," Applied Biological Research, vol. 2, no. 1-2, pp. 39–44, 2000.

50. H. N. Ravankar, R. Patil, and R. B. Puranik, "Decomposition of different organic residues in soil," PKV Research Journal, vol. 24, no. 1, pp. 23–25, 2000.

51. K. L. Sharma, K. Srinivas, U. K. Mandal et al., "Kinetics of decomposition of un-conventional farm-based crop residues and their composting and quality monitoring," Communications in Soil Science and Plant Analysis, vol. 38, no. 17-18, pp. 2423–2444, 2007.

52. Yadvinder-Singh, Bijay-Singh, and J. Timsina, "Crop residue management for nutrient cycling and improving soil productivity in rice-based cropping systems in the tropics," Advances in Agronomy, vol. 85, pp. 269–407, 2005.

53. R. Ordonez-Fernandez, A. Rodríguez-Lizana, R. Carbonell, P. González, and F. Perea, "Dynamics of residue decomposition in the field in a dryland rotation under Mediterranean climate conditions in southern Spain," Nutrient Cycling in Agroecosystems, vol. 79, no. 3, pp. 243–253, 2007.

54. M. Corbeels, G. Hofman, and O. van Cleemput, "Nitrogen cycling associated with the decomposition of sunflower stalks and wheat straw in a Vertisol," Plant and Soil, vol. 218, no. 1-2, pp. 71–82, 2000.

55. M. Jalali, "Phosphorus availability as influenced by organic residues in five calcareous soils," Compost Science and Utilization, vol. 17, no. 4, pp. 241–246, 2009.

56. S. K. Das, K. L. Sharma, K. Srinivas, M. N. Reddy, and O. Singh, "Phosphorus and sulphur availability in soil following incorporation of various organic residues," Journal of the Indian Society of Soil Science, vol. 43, no. 2, pp. 223–228, 1995.

57. D. K. Das and A. M. Puste, "Influence of different organic waste materials on the transformation of nitrogen in soils," TheScientificWorldJOURNAL, vol. 1, pp. 658–663, 2001.

58. V. U. Raut, R. T. Bhowate, and A. G. Waghmare, "Effect of crop residues on nutrient contents in greengram-sunflower cropping sequence," Green Farming, vol. 1, no. 1, pp. 14–19, 2010.

59. R. Santamaría, M. E. Aguirre, and M. A. Commegna, "Effects of sunflowers hull incorporation on the distribution of water extractable elements in the soil," Agrochimica, vol. 52, no. 4, pp. 243–252, 2008.

60. M. Jalali and F. Ranjbar, "Rates of decomposition and phosphorus release from organic residues related to residue composition," Journal of Plant Nutrition and Soil Science, vol. 172, no. 3, pp. 353–359, 2009.

61. Z. Gong, P. Li, B. M. Wilke, and K. Alef, "Effects of vegetable oil residue after soil extraction on physical-chemical properties of sandy soil and plant growth," Journal of Environmental Sciences, vol. 20, no. 12, pp. 1458–1462, 2008.

62. K. Prasad, R. Pyare, C. P. Verma, and S. Chaudhary, "Effect of soil conditioners and fertilizer application on soil properties with maize residue incorporation under maize-wheat sequence," Plant Archives, vol. 5, no. 2, pp. 421–427, 2005.

63. A. Scagnozzi, A. Saviozzi, R. Levi-Minzi, and R. Riffaldi, "Nutrient release from decomposing crop residues in soil: a laboratory experiment," The American Journal of Alternative Agriculture, vol. 12, no. 1, pp. 10–13, 1997.

64. K. I. Paul, A. S. Black, and M. K. Conyers, "Effect of plant residue return on the development of surface soil pH gradients," Biology and Fertility of Soils, vol. 33, no. 1, pp. 75–82, 2001.

Soil Health Management under Hill Agroecosystem of North East India

R. Saha, R. S. Chaudhary, and J. Somasundaram

Indian Institute of Soil Science, Indian Council of Agricultural Research, Nabibagh, Berasia Road, Bhopal, Madhya Pradesh 462 038, India

ABSTRACT

The deterioration of soil quality/health is the combined result of soil fertility, biological degradation (decline of organic matter, biomass C, decrease in activity and diversity of soil fauna), increase in erodibility, acidity, and salinity, and exposure of compact subsoil of poor physicochemical properties. Northeast India is characterized by high soil acidity/Al^{+3} toxicity, heavy soil, and carbon loss, severe water scarcity during most parts of year though it is known as high rainfall area. The extent of soil and nutrient transfer, causing environmental degradation in North eastern India, has been estimated to be about 601 million tons of soil, and 685.8, 99.8, 511.1, 22.6, 14.0, 57.1, and 43.0 thousand tons of N, P, K, Mn, Zn, Ca, and Mg, respectively. Excessive deforestation coupled with shifting cultivation practices have resulted in tremendous soil loss (200 t/ha/yr), poor soil physical health in this region. Studies on soil erodibility characteristics under various land use systems in Northeastern Hill (NEH) Region depicted that shifting cultivation had the highest erosion ratio (12.46) and soil loss (30.2–170.2 t/ha/yr), followed by conventional agriculture system (10.42 and 5.10–68.20 t/ha/yr, resp.). The challenge before us is to maintain equilibrium between resources and their use to have a stable ecosystem. Agroforestry systems like agri-horti-silvi-pastoral system performed better over shifting cultivation in terms of improvement in soil organic carbon; SOC (44.8%), mean weight diameter; MWD (29.4%), dispersion ratio (52.9%), soil loss (99.3%), soil erosion ratio (45.9%), and in-situ soil moisture conservation (20.6%) under the high rainfall, moderate to steep slopes, and shallow soil depth conditions. Multipurpose trees (MPTs) also played an important role on soil rejuvenation. Michelia oblonga is reported to be a better choice as bioameliorant for these soils as continuous leaf litter and root exudates improved soil physical behaviour and SOC considerably. Considering the present level of resource degradation, some resource conservation techniques like zero tillage/minimum tillage, hedge crop, mulching, cover crop need due attention for building up of organic matter status for sustaining soil health.

INTRODUCTION

Soil degradation has raised some serious debate, and it is an important issue in the modern era. It refers to the decline in soil's inherent capacity to produce economic goods and perform ecologic functions. It is the net result of dynamic soil derivative and restorative processes regulated by natural and anthropogenic factors. The degree of soil degradation depends on soil's susceptibility to degradative processes, land use, the duration of degradative land use, and the management. Soil and water degradation are also related to overall environmental quality, of which water pollution and the "greenhouse effect" are two major concerns of global significance. Recent global concerns over increased atmospheric CO_2, which can potentially alter the earth's climate systems, have resulted in raising interest in studying Soil organic matter (SOM) dynamics and carbon (SOC) sequestration capacity in various ecosystems [1]. Soils represent an important terrestrial stock of C and approximately two to three times as much as terrestrial vegetation and atmosphere, respectively, and the C in the SOM of agricultural land is composed of dominant terrestrial C stock. Soil quality is the capacity of a soil to function within ecosystem boundaries to sustain biological productivity, maintain environmental quality, and promotes plant and animal health and thus has a profound effect on the health and productivity of a given ecosystem and the environment related to it.

The North Eastern parts of India, comprising the states of Arunachal Pradesh, Assam, Manipur, Meghalaya, Mizoram, Nagaland, Sikkim, and Tripura, lies between 22°05' and 29°30' N latitudes and 87°55' and 97°24' E longitudes. The region is characterised by diverse agroclimatic and geographical situations. About 54.1 percent of the total geographical area is under forests, 16.6 percent under crops, and the rest either under nonagricultural uses or uncultivated land. The low area under agricultural crops is due to natural corollary of the physiographic features of the region, as major chunk of the land has more than 15 percent slope, undulating topography, highly eroded and degraded soils, and inaccessible terrain. Continuous dilution of the forest cover in the region due to shifting cultivation, firewood, and timber collection is posing the most crucial problem resulting in poor soil health and environmental degradation in the hills.

SHIFTING CULTIVATION

Shifting cultivation, also known as Jhum cultivation, is the most traditional and dominant land use system in this region. On an average, 3,869 km^2 areas is put under shifting cultivation every year. Shifting cultivation in its more traditional and cultural integrated form is an ecological and economically viable system of agriculture as long as population densities are low and jhum cycles are long enough to maintain soil fertility. The system involves cultivation of crops in steep slopes. Land is cleared by cutting of forests, bushes, and so forth up to the stump level, leaving the cut materials for drying and finally burning to make the land ready for sowing of seeds of different crops before the onset of rains. The cultivation is confined to a village boundary and often after two or three years, the cultivated area is abandoned and a new site is selected to repeat the process. The shifting cultivation became unsustainable today primarily due to the increase in population that led to increase in food demand. Jhuming cycle in the same land, which extended to 20–30 years in earlier days, has now been reduced to 3–6 years [2]. Land degradation in the region is 36.64% of the total geographical area, which is almost double than the national average of 20.17% [3]. The problem of land degradation is much serious in the states like Manipur, Nagaland, and Sikkim, where more than 50% of total geographical area is defined as wastelands. Of various degradation types, water erosion, reduced infiltration, acidification, nutrient leaching, burning of vegetation, decline in vegetative cover, and biodiversity are important in context to the NE region.

EFFECT OF SHIFTING CULTIVATION

Change in Forest Cover

The total forest cover in the region is 1, 41,652 km^2, which is about 54.1% of the geographic area as against the national average of 19.39%. [4]. Manipur and Meghalaya have dense forest cover of 25.57 and 25.33%, respectively (Table 1). Similarly for Nagaland, Sikkim, Tripura, and Mizoram, the dense forest cover is 32.53, 33.70, 33.02 and 42.39%, respectively. Among seven sisters of NEH, Arunachal Pradesh is the only state, which has the dense forest cover of 64.0%. Since shifting cultivation is still practiced in the region, and every year dense forest is converted into jhum fields, there is drastic reduction in dense forest cover (canopy density > 40%) in most of the states.

Table 1: Trends of forest loss/gain (km²) in NEH region

States	1999 Assessment			2001 Assessment			Net difference
	Dense forest cover	Open forest cover	Total forest cover	Dense forest cover	Open forest cover	Total forest cover	
Arunachal Pradesh	57,756	11,091	68,847	53,932	14,113	68,045	(−) 802
*Assam	15,548	8,276	23,824	14,517	9,171	23,688	(−)136
Manipur	5,936	11,448	17,384	5,710	11,216	16,926	(−) 458
Meghalaya	5,925	9,708	15,633	5,681	9,903	15,584	(−) 49
Mizoram	3,786	14,552	18,338	8,936	8,558	17,494	(−) 844
Nagaland	5,137	9,027	14,164	5,393	7,952	13,345	(−) 819
Sikkim	2,363	755	3,118	2,391	802	3,193	75
Tripura	2,228	3,517	5,745	3,463	3,602	7,065	1320
**Total	83,131	60,098	1,43,229	85,506	56,146	1,41,652	(−) 1577

*Data for Assam is during the assessment year of 1997–1999 and **total reports only for NEH region.

Source: [4].

Effect of Burning on Soil Fertility

The burning process related to shifting cultivation practices has tremendous effect on soil ecosystem. The impact of fire on ecosystem is profound and its consequences are dependent on intensity and frequency of fire, proportion of biomass burned, the time of monsoon setting, and total annual precipitation. The extent to which organic matter is transformed into ash depends on a number of factors viz intensity and duration of fire, fuel load, moisture content in the fuel, weather, and topography. Burning of above-ground vegetation showed an increase in pH and cations and a decrease in carbon and nitrogen contents in the surface soil [5]. Quick release of nutrients especially cations after burning has been reported by Kellman et al. [6]. The organic carbon content of soil decreased drastically after burning because of oxidation loss. Rise in pH, temperature, and bases of the soil might have increased the microbial activity after burning which in turn resulted in accelerating mineralization of organic N to inorganic forms [7, 8].

Soil Erosion and Nutrient Loss

Soil erosion under shifting cultivation is highly erratic from year to year depending on rainfall characteristics. Studies on steep slopes (44–53%) have indicated the soil loss to the tune of 40.9 t/ha and the corresponding nutrient losses per hectare are 702.9 kg of organic carbon, 63.5 kg of P and 5.9 kg of K [9]. The soil loss from hill slopes (60–79%) under first year, second year, and abandoned jhum was estimated to be 147, 170, and 30 t/ha/yr [10]. In general, tolerable soil loss (T) value is 11.2 Mg/ha/yr (5.0 t/ac/yr) while it is between 5.0 and 12.5 Mg/ha/yr (2.2 and 5.6 t/ac/yr) in North West Himalayas [11]. During first few years of clearing, carbon and nitrogen levels decrease rapidly. According to one estimate annual loss of top soil, N, P and K due to shifting cultivation is 88346, 10669, 0.372, and 6051 thousand tones in the region [12]. Singh et al. [13] reported nutrient loss to the tune of 6.0 million tons of organic carbon, 9.7 tones of available P, and 5690 tones of K from the NEH region. Nutrient losses from the jhum field through runoff and percolation are rather heavy during cropping.

LONG-TERM STRATEGIES FOR RESOURCE CONSERVATION AND IMPROVEMENT IN SOIL HEALTH

Nearly 37.1% of the total geographical area in Northeast India is under the threat of land degradation, where erosion is a major land degradative process. With the great concern of poor soil health and severe land degradation, there is a need of viable option for ecorestoration and maintenance of soil resources which could sustain long-term soil productivity and improve food security of the poor tribal farmers of northeast India under the humid subtropical climate of the north-eastern Himalayan region. Three broad strategies, suitable for different land situation, elevation, and topography prevailing in this region, are discussed here.

Multipurpose Trees (MPTs)

The multipurpose tree species (MPTs) form an integral component of different agroforestry interventions in crop sustainability. The MPTs, besides furnishing the multiple outputs like fuel, fodder, timber, and other miscellaneous products, help in improvement of soil health and other ecological conditions. Farmers of the region integrate various tree species in different land use in the region; however, priority species vary from state to state and even from place to place within a state based on ethnic diversity and food habits of the tribal communities. In the region, as many as 40 promising species are cultivated in tropical and subtropical region, and 30 in temperate zone of the region in different farming systems. Besides, 28 bamboo species and 2 genera of cane also find a place in various agroforestry programmes. Tree density ha^{-1} is also a crucial factor on sloppy lands. In general, optimum tree density in case of agri-horticulture system is 400 trees/ha, while in agri-silviculture, it is 200 plants/ha so as to minimize the effect of shade and biochemical interactions on growth and production of agricultural crops [14, 15].

Long-term effect of various multipurpose tree species on soil physical behaviour has been studied [16]. Multipurpose tree species with greater surface cover, constant leaf litter fall, and extensive root system increased soil organic C by 96.2%, porosity by 10.9%, aggregate stability by 24.0%, and available soil moisture by 33.2% and simultaneously reduced bulk density and erosion ratio by 15.9 and 39.5%, respectively (Table 2). Among the tree species tested, P. kesiya, M. oblonga and Alnus nepalensis were found suitable as bioameliorant in hilly terrain of northeast India in terms of organic matter buildup through presence of leaf litter, better soil aggregation, transmissivity, and infiltrability through extensive root system, improved soil conservation through constant surface cover with leaf biomass. Such improvement in soil hydrophysical properties in tree-based system has a direct bearing on long-term sustainability, productivity, and soil quality in hilly ecosystem.

Table 2: Effect of various MPTs on soil physical properties

MPTs	Organic C (g kg−1)	Aggregate stability	Available water (m3 m−3)	Infiltration rate (mm h−1)	Erosion ratio
Pinus kesiya	35.4	75.6	0.220	8.04	0.20
Alnus nepalensis	32.2	72.1	0.201	7.28	0.23
Parkia roxburghii	23.1	63.4	0.192	4.85	0.30
Michelia oblonga	33.6	73.2	0.210	6.10	0.22
Gmelina arboria	28.6	67.9	0.183	5.36	0.24
Control (No tree)	15.6	56.8	0.151	3.84	0.39

Source: [16].

Agroforestry Interventions in Degraded Lands

The region has a very high rate of land degradation. In this region, 7.85 million ha area is degraded which need rehabilitation through various agroforestry models [3]. Agroforestry system (AFS) has today become an established approach of integrated land management system not only for renewable resource production but also for ecological consideration. It represents the integration of agriculture and forestry to increase the productivity and sustainability of farming system.

Soil Fertility Buildup

Study revealed that organic carbon, available P, and exchangeable cations contents in surface soil ranged in between 2.0–2.5%, 10.4–13.2 ppm, and 5.9–8.4 cmol (p^+) kg^{-1}, respectively, under jackfruit-based AFS, while 1.5–1.8% organic carbon, 3.8–6.7 ppm available P, and 3.9–5.9 cmol (p^+) kg^{-1} total cations were found under arecanut/khasi mandarin-based AFS [17].

In an another study, long-term effect of agri-horticulture (comprising Khasi mandarin + agricultural crops, and Assam lemon + agricultural crops), agri-silviculture (multipurpose tree species + annual agricultural crops), silvi-horti-pastoral (alder + pine apple + fodder grasses), and multistoried AFS (alder + tea + black pepper + annual agricultural crops between the tree rows) on soil properties and fertility status was evaluated in acid Alfisol of Meghalaya compared with natural forest as a control. In all the AFS, significant (1.17–1.65 fold) increase in organic carbon was found as compared to initial status, the maximum contribution being by silvi-horti-pastoral AFS. The same system also registered 43.2% higher exchangeable Al compared to natural forest and consequently a maximum decrease of 0.50 units in pH (Table 3). The exchangeable Ca, Mg, Na, K, and Al and available N, P, and K content were higher in all the systems compared to natural forest and the content of these nutrients decreased with increasing soil depth [18].

Table 3: Effect of agroforestry systems on soil properties

Soil properties	Agroforestry systems					Natural forest
	Agrisilviculture	Agrihorti culture (khasi mandarin + crops)	Agrihorti culture (Assam lemon + crops)	Silvihorti pastoral (Alder + pine apple + fodder grass)	Multistoried AFS (Alder + tea + black pepper + crops)	
pH	4.65	4.62	4.80	4.25	4.61	4.62
Organic C (%)	1.62	1.55	2.02	2.19	1.91	1.92
Exchangeable Ca [cmol (p+) kg−1]	0.40	0.86	0.74	0.31	0.65	0.26
Exchangeable Mg [cmol (p+) kg−1]	0.75	0.51	0.33	0.48	0.71	0.16
Exchangeable K [cmol (p+) kg−1]	0.232	0.244	0.249	0.238	0.201	0.169

Exchangeable Na [cmol (p+) kg−1]	0.201		0.220	0.194	0.195	0.197	0.196
Exchangeable Al [cmol (p+) kg−1]	2.65		2.70	2.20	3.15	2.05	2.20
Available N (ppm)	190.1		180.8	203.6	199.4	216.9	167.2
Available P (ppm)	2.75		4.10	5.36	0.94	3.36	0.63
Available Fe (ppm)	8.9		10.4	12.8	10.9	13.9	7.3
Available Mn (ppm)	0.58		0.92	0.79	0.83	1.04	0.04
Available Zn (ppm)	0.08		0.05	0.07	0.006	0.08	0.025
Available Cu (ppm)	0.21		0.23	0.37	0.30	0.27	0.10

Source: [18]

In a study under Farming System Research Project (FSRP) carried out at ICAR Research Complex, Barapani, effect of various AFS like silvi-pastoral, silvi-horticulture, agri-horti-silvipastoral has been evaluated [19] after 17 years of their adoption on soil fertility indices (Table 4). The natural fallow and abandoned jhum land at Umiam were taken for comparison. Organic carbon content increased in all the AFS including natural fallow, however, the quantity largely depended on the nature of vegetation in different systems. Adoption of different cropping pattern in various AFS markedly influenced the exchangeable Ca, Mg, and K content in the soil. Maximum accumulation of these cations was recorded under agri-horti-silvipastoral and silvi-horticulture AFS followed by natural fallow and silvi-pastoral systems. Accumulation of exchangeable K was maximum in silvi-horticulture followed by agri-horti-silvipastoral. The available N, P, and S were higher in agri-horti-silvipastoral and silvi-horticulture compared to natural fallow and silvi-pastoral AFS.

Table 4: Effect of different land use systems developed under FSRP, Meghalaya on soil properties

Characteristics	Agroforestry systems					
	Natural fallow	Abandoned jhum land	Silvipastoral	Agri-horti-silvipastoral	Silvihorti-culture
pH	4.99 (4.90)	4.76 (5.20)	4.52 (5.10)	4.92 (4.90)	4.91 (4.90)
Organic C (%)	2.94 (1.85)	3.42 (1.90)	2.61 (1.80)	2.97 (1.82)	2.97 (1.80)
Exchangeable Ca [cmol (p+) kg−1]	1.96 (1.15)	1.57 (1.16)	1.25 (1.10)	2.11 (1.20)	2.00 (1.20)
Exchangeable Mg [cmol (p+) kg−1]	0.55 (1.15)	0.38 (1.16)	0.43 (1.20)	1.45 (1.20)	0.85 (0.60)
Exchangeable Al [cmol (p+) kg−1]	0.88	1.30	1.56	0.90	0.90
Available N (ppm)	179.2	251.1	214.5	220.3	210.9
Available P (ppm)	1.9	2.0	2.1	16.6	12.9
Available K (ppm)	175.6	130.8	98.0	162.7	265.0
Available S (ppm)	14.5	14.8	10.4	19.9	12.9

Figures in parentheses indicate the initial values at the start of the project.

Source: [19].

Soil Physical Health

Effect of various land use systems on soil physical properties shown in Table 5 indicated that the maximum reduction in bulk density over shifting cultivation was recorded in forest (17.6%) followed by agri-horti-silvi-pastoral (14.3%), livestock based (13.4%), natural fallow, and agriculture system (12.6%). Higher percentage of macroaggregates (54.5%), organic C content (2.95%), and biotic activity were also observed in forest ecosystem. Soil biota influences soil properties through formation of stable aggregates, development of organomineral complexes by improving macroporosity and continuity of pores from surface to the subsoil which ultimately increase the water transmission and reduce run-off. Higher transmission and storage pore volume coupled with lower value of residual pores associated with modified land use systems as compared to shifting-cultivated plots was thus an indication of maintaining the pore geometry of the soil under these systems. The better soil aggregation under natural forest, multistoried AFS, and silvihortipastoral systems maintaining intensive vegetative cover throughout the year could be ascribed to the effect of higher percentage of organic matter, clay content, and high amount of Al and Fe oxides in soil.

Table 5: Effect of different land use systems developed under FSRP, Meghalaya on soil physical properties

Soil properties	Land use systems						
	Agriculture	Agri-horti-silvipastoral	Forestry	Livestock based	Natural fallow	Shifting cultivation
Bulk density (Mg m−3)	1.04	1.02	0.98	1.03	1.04	1.19
Total Porosity (%)	59.67	60.47	62.02	60.08	66.23	53.88
Macroaggregates (>0.25 mm)	21.72	54.19	54.47	50.02	50.53	18.17
Microaggregates (<0.25 mm)	47.85	23.23	23.81	22.80	21.90	42.34
MWD (mm)	2.76	2.99	3.16	2.85	2.93	2.31
Available water (m3 m−3)	0.210	0.222	0.231	0.220	0.233	0.169
Hydraulic Conductivity (cm hr−1)	2.74	4.72	5.47	2.95	6.66	2.09

Source: [20].

Soil and Water Conservation

Some of the potential farming systems such as agriculture on bench terraces, horticulture, and agri-horti-silvipastoral systems have been evaluated [21] at the experimental watershed of ICAR Research Complex at Barapani for long-term runoff, soil and nutrient losses, production behaviour, biotic and abiotic changes, and so on. The data indicated that mixed land use systems with appropriate soil conservation measures, namely, bench terraces, contour trenches, and so forth, were the most effective in retaining 90–100% annual rainfall and simulated the effects of natural forest. The contributions to stream flow in the watersheds having substantial area under natural forest is primarily by subsurface flow (base flow). The watersheds having continuous stream flow characteristics generated base flow to the extent of 70–90% of its total water yields. As expected, the watershed treated with jhum (shifting) cultivation yielded the highest peak runoff while the one left undisturbed with natural vegetation gave the minimum peak runoff. The results revealed that agroforestry and other mixed land use systems most effectively conserved moisture and substantially reduced peak runoff (Tables 6(a) and 6(b)). The low erosion ratio values in silvi-horti-pastoral and multistoried AFS (3.07 and 3.06, resp.) showed that these systems were the most suitable for soil and water conservation in hilly ecosystem [22]. This could be ascribed to the effect of heavy litter fall, which might have increased the cohesiveness in the soil system after decomposition and also binds the soil tightly in lower horizons by their deep root systems.

Table 6: (a) Pretreatment (year) precipitation, storm flow, peak flow rate in different land use systems. (b) posttreatment water yield, base flow, and peak flow in different land use systems (average of nine years)

(a)

Land use system	Precipitation (mm)	Threshold rainfall (mm)	Total water yield (mm)	Total water yield (% of rainfall)	Surface runoff (mm)	Peak flow (mm hr−1)
Dairy based farming	2249.30	363.20	27.21	1.20	27.21	3909

Forestry block	2249.30	399.90	655.21	29.12	54.30	16.94
Agroforestry	2249.30	533.70	32.55	1.45	9.90	6.17
Agropastoral	2249.30	364.30	25.50	1.13	25.50	31.86
Agrohortisilvipastoral	2249.30	348.60	4.10	0.18	4.10	10.45
Natural fallow	2249.30	541.70	2.87	0.13	2.87	13.65
Shifting cultivation	2249.30	1634.5	15.88	0.70	15.88	35.30

(b)

Land use systems	Annual water yield range (mm)	Mean water yield (mm)	Mean water yield (% of annual rainfall)	Maximum peak flow (mm hr−1)
Dairy based farming	0–66.699	9.56	0.37	7.81
Forestry block	67.42–1013.88	371.90	4.73	13.54
Agroforestry	39.31–648.26	241.14	9.55	12.87
Agropastoral	0.60–62.49	12.47	0.69	20.71
Agrohortisilvipastoral	0.24–121.91	28.98	1.14	12.07
Natural fallow	0–51.39	11.77	0.46	4.49
Shifting cultivation	0–517.72	102.94	4.07	86.10

Source: [21].

Soil C Sequestration Potential

Assessment of soil quality is an invaluable tool in determining the sustainability and environmental impact of agricultural ecosystems. Soil quality under different agroecosystems using soil organic carbon (SOC) and soil microbial C (SMBC) as soil quality indicators suggests that the shifting cultivated areas had the lowest SMBC value of 192 mg/kg while soil under Michelia oblonga plantation had the significantly () highest value of 478 mg/kg. The proportion of SMBC to total soil organic carbon (SOC) was in the range of 0.76 to 1.96% across all the systems. Multipurpose tree species like P. kesiya, A. nepalensis, P. roxburghii, M. oblonga, and G. arboria with greater surface cover, constant leaf litter fall, and extensive root systems increased soil organic carbon by 96.2% (Table 7), helped with better aggregate stability by 24.0%, improved available soil moisture by 33.2%, and in turn reduced soil erosion by 39.5% [16, 23]. Similarly, a comparative study on the effect of various MPTs on soil organic carbon pool (Table 8) showed a concomitant rise in SOC in soils under MPTs and a subsequent decline in soils of open space over 4–16 years. Maximum rise in SOC was noticed in soils of A. indica (28.6 Mg/hm^2) followed by A. Aurculiformisi (21.9 Mg/hm^2), G. arborea (21.8 Mg/hm^2), M. Champaca (16.7 Mg/hm^2), and so forth. The minimum rise in SOC was noted in soils under T. grandis. So an increase of SOC was noted from 3.8 Mg/hm^2 in soils of open space to 19.5 Mg/hm^2 in that under MPTs after 16 years. The comparatively high humin carbon present in soils under A. auriculiformis, L. leucocephala,and G. Arborea indicated the enhanced storage of organic carbon pool in agroforestry systems [24]. Swamy et al. [25] estimated that a six-year-old G. arborea, based agri-sivicultural systems in India sequestered 31.4 Mg hm^{-2} carbon.

Table 7: Growth, litter production, fine root biomass of promising MPTs in humid tropics, and their contribution on SOC content

MPT	Annual litter production (g m−2)	Time required for decomposition (days)	Total fine root biomass (g m−2)	Organic C (g kg−1)
P. kesiya	621.5	718	496.75	35.4
A. nepalensis	473.75	350	435.50	32.2
P. roxburghii	341.75	385	415.50	23.1
M. oblonga	512.25	390	462.00	33.6
G. arboria	431.75	360	419.00	28.6

Source: [16].

Table 8: Changes in SOC (Mg hm^{-2}) over the years under various MPTs in humid tropics MPTsYears481216

MPTs	Years			
	4	8	12	16
A. auriculiformis	11.1	11.9	17.9	21.9
M. alba	9.9	9.9	9.9	15.9
L. leucocephala	11.5	11.5	12.8	16.7
D. sissoo	13.1	12.5	13.1	13.9
G. maculate	13.1	13.1	13.9	14.9
A. indica	10.9	10.9	14.7	28.6
M. champaca	13.9	13.7	13.9	16.9
E. hybrid	9.9	9.9	14.9	16.1
T. grandis	11.5	11.3	11.5	12.9
G. arborea	12.2	12.2	12.8	21.8
S. saman	10.6	11.3	11.3	13.9
A. procera	13.5	13.1	13.5	14.7
Open space (Control)	11.9	11.9	11.1	9.1

Resource Conservation Techniques

Conservation Tillage

Conservation tillage are system of managing crop residue on the soil surface with minimum disturbance. The stubble mulch or reduced tillage/minimum tillage, no tillage and direct drill are components of conservation tillage. The objectives are (i) to leave enough plant residues on the soil surface at all times for water, and wind erosion control, (ii) to conserve soil and water and (iii) to reduce energy use [26]. Some of the conservation tillage practices followed in hill ecosystems are discussed here.

In-Situ Residue Management

Low native soil nitrogen (N) and very low phosphorus (P) coupled with apathy of farmers towards use of fertilizer is the major constraints limiting the rice productivity in NEH Region of India. Productivity and nutrient recycling potential in rice (Oryza sativa L.)—vegetables cropping sequences under low input in-situ residue management under rainfed condition was evaluated on lowland situation at ICAR Research Complex for NEH Region, Umiam, Meghalaya. After harvesting of rice, five vegetable crops, viz., tomato, potato, frenchbean, cabbage, and carrot, were grown. No external input including fertilizer, pesticides, and so forth was applied except one hand weeding at 30 days after transplanting in case of rice. In case of vegetables, only one earthing up and intercultural operations were done as per the requirement of the crops. Only the economic parts of crops were harvested and left-out portion including weed residues were chopped and incorporated into the soil. A considerable amount of nutrients were recycled through in-situ weed biomass incorporation. The weed biomass ranged from 37.5 q/ha with rice-tomato to 50.6 q/ha in rice-fallow. Highest amount of NPK recycling was recorded from rice-potato sequence. Soil fertility in terms of available NPK status analysed after 4 years was found stable in all the crop sequences except rice-cabbage, where it declined slightly. The soil biological properties like population of Rhizobium, bacteria, phosphorus solubilizing microorganisms, and earthworm activity all were found remarkably higher in experimental field compared to plots that are managed inorganically.

Incorporation of Jungle Grass

Long-term effects of different locally available grasses and weeds on soil hydro-physical properties and rice yield through a 5-year field experimentation under hilly ecosystem of Meghalaya depicted that incorporation of jungle grass (Ambrosia spp.), in puddled rice soil improved soil organic carbon (SOC) by 21.1%, the stability of microaggregates, moisture retention capacity, and infiltration rate of the soil by 82.5, 10, and 31.3%, respectively, and soil bulk density decreased by 12.6% [27]. Locally available jungle grasses are equally good as an organic amendment, which would also ease the problem of disposal of these grasses during peak monsoon. Therefore, these organic sources may serve as alternative to farm yard manure (FYM) and have a dramatic effect on long-term productivity of rice.

Zero Tillage

Zero tillage in rice-based system improves physical properties of soil like soil structure, increase the relative proportion of biochannels, macropores, and decrease the susceptibility of crusting. It has been observed that the bulk density of soil decreased about 25%, total porosity and soil aggregates increased by 29 and 32%, respectively, over the conventional tillage practices (2-3 passes of powertiller/spade). It also increases the SOC content by 12.5%, available P by 14.3%, and K by 29.4% over conventional tillage. Zero tillage saved 20% energy (Figure 1) and fertilizer needs as compared other conventional tillage methods by conserving soil and water [28] without jeopardizing the crop production (rice yield of 37 q/ha). In other tillage practices like power-tilled, desiploughed, or manually weeding, the energy in terms of labour requirement was much higher.

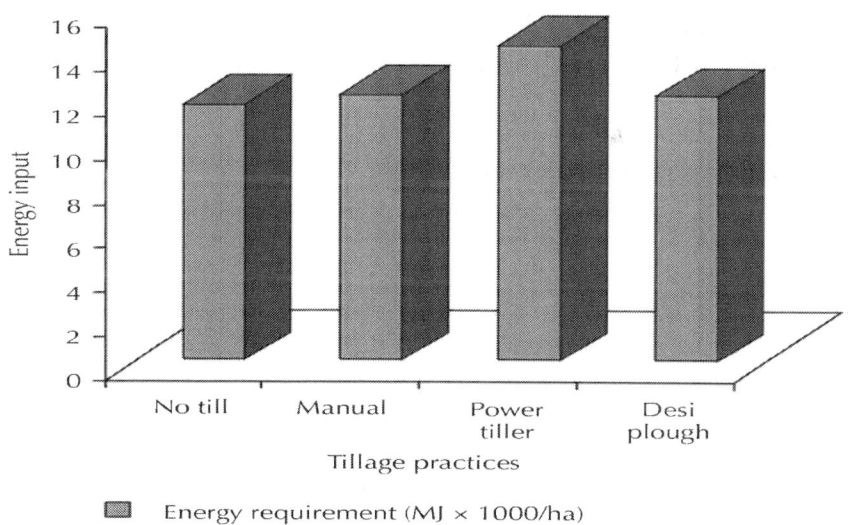

Figure 1: Energy requirement of different tillage practices. Source: [28].

Integrated Plant Nutrient Supply

Integrated use of balanced inorganic fertilizer in combination with lime and organic manure sustains a better soil health for achieving higher crop productivity under intensive cropping systems in hilly ecosystem of north eastern India. Study suggests that addition of NPK fertilizers along with organic manure, lime, and biofertilizers had increased SOC content, aggregate stability, moisture retention capacity, and infiltration rate of the soil while reducing bulk density. The SOC content under the treatment 100% NPK + lime + biofertilizer + FYM was significantly higher (68.6%) than control plots [29].

Pastural Development

Resource conserving and environmental friendly production strategies are desirable for agrarian economies. Grass cover is the key factor in improving soil physicochemical health by assuring regular addition of organic matter, thus reducing surface runoff and soil erosion. Some promising perennial grasses like Setaria, Congosignal, Guinea, Napier, and Broom grass were tested for their effect on soil physicochemical properties. Study [30] revealed that continuous 15 years grass covers significantly increased the SOC, the highest SOC content with Setaria (2.24%). Similarly, Soil microbial biomass carbon, soil aggregation, and infiltration rate under various grass covers were also high as compared to plots without grass covers.

Hedgerow Intercropping

As the trees have long gestation period, farmers may be reluctant to cultivate the trees mainly due to prevailing land tenure system in the region. However, cultivating various hedgerow species even injhum field could be better option for them as these species have short gestation period. Hedgerows alone reduced soil loss by 94% and run-off by 78%. When twigs and tender stem of hedge plants are used for mulch, it conserved 83% of the soil and 42% of rainfall. In a study conducted at Changki, Nagaland in NEH region, the soil loss was reduced by 22% with the incorporation of hedgerow species in the jhum fields compared to traditional jhum site (38.14 t/ha/yr). Thus contour hedgerow technology provides an option for farming on the hill slopes on a sustainable basis. Growing of nitrogen fixing hedge species on the field bunds helps in fixation of atmospheric nitrogen and reduces the leaching losses of mineral nitrogen. Their vigorous root system mobilizes phosphorus, potassium, and other trace elements. ICAR Research Complex for NEH Region has also screened various hedgerow species for plantation, and Cajanus cajan, Crotalaria tetragona, Desmodium rensonii, Flemingia macrophylla, Indigofera tinctoria, Tephrosia candida, and Gliricidia maculata have been found suitable for farming in Eastern Himalayas. Survival percentage of these species ranged from 60.0 to 80.0 over degraded sites. The total N, P, and K concentration in the foliage of hedgerow species ranged from 3.23–3.86; 0.32–0.81; 1.26–1.67%, respectively. Total leaf biomass production on the dry weight basis after one year of growth was found to be highest in C. tetragona (22.98 q/ha) followed by G. maculata (20.75 q/ha), I. tinctoria (16.99 q/ha), and T. candida (15.30 q/ha). Among the hedgerow species, C. tetragona enriched the soil fertility more efficiently as it accumulated higher amount of total N, P, and K (79.74, 11.03, and 37.46 kg/ha) through its leaf incorporation. The recycling of bases in litter of hedgerow could potentially counteract the acidification [31]. The incorporation of leaf biomass of T. candida improved the pH in acid soil by increasing 0.49 units from the initial level at surface soil. Thus, the biomass produced from hedgerows showed a favorable influence on soil acidity.

Organic Farming

Organic farming is primarily in operation in areas under shifting cultivation and traditional land use systems in north east India. Nearly 57.1% of total geographical area (TGA) in India is under the threat of land degradation mainly by water erosion. On an average, 37.1% of TGA in NE India is in degraded state. Fertilizer use in most of the states of the region is far below the national average. The use of N, P, and K through fertilizer in the region is only 13.37, 11.12, and 11.0% of the crop removal thus necessasiating the organic source of nutrition in the domain of soil health management. Organic sources if pooled together can supply 13.07 kg N/ha, 7.18 kg phosphate/ha, and 7.34 kg potash/ha in NE India. The micronutrient supply from organic sources may be adequate. Substantial amount of potash can be obtained from crop residues if managed to add in soils. Biofertilizers in case of adequate supply can produce an increase (5–30%) in yield. Vermicomposting of rural wastes holds a great promise in mitigating nutrient hunger of soils in NE India considering supply of composting earthworms and need based training in compost technology. Soil amelioration with the use of limestone deposit available in north east can be brought in use. Finally, watershed based technology with proper soil and water conservation measures can be an effective avenue to nurture soil health for sustainable organic food production.

EPILOGUE

Even today, Jhuming is considered as a major source of rural economy in north eastern part of India and will remain as important one as it is associated with socioeconomic and cultural systems of the people of this region. Because of this, degradation will continue in the years to come and may reach to the extent of out of control, if proper care is not taken right now. Therefore, to reduce all types of degradation level, a comprehensive forest policy is required as a long-term strategy in the region for sustainability and augmentation of food, fuel, fodder, and timer requirements. In this direction, agroforestry coupled with some sound resource conservation techniques needs to be strengthened for long-term sustainable production and environmental conservations in fragile ecosystem which will contribute to improved food security and income generation for resource poor farmers and protect the environments.

Integrated farming system (IFS) has emerged as a well-accepted, single window, and sound strategy for harmonizing simultaneously jointmanagement of land, water, vegetation, livestock, and human resources, The IFS developed for hill areas could reduce the risk of soil degradation, produce the soils productive potential, and reduce the risks of environmental degradation. Besides, these interventions having a tree crop with a high quality of leaf litter and root binding ability reduce erodibility of rainfall/runoff and improve the physicochemical conditions. Attempt should also be made to manage soil health through addition of organic inputs in this region.

REFERENCES

1. W. H. Schlesinger, "Carbon sequestration in soils," Science, vol. 284, no. 5423, p. 2095, 1999.

2. D. N. Borthakur, Agriculture of the North Eastern Region with Special Reference to Hill Agriculture, Beecee Prakashan, Guwahati, India, 1992.

3. Anonymous, Wastelands Atlas of India, Ministry of Rural Development, Government of India and National Remote Sensing Agency, Hyderabad, India, 2000.

4. Anonymous, State of Forest Report, Forest Survey of India. Ministry of Environment and Forests, Government of India, 2001.

5. S. C. Ram and P. S. Ramakrishnan, "Hydrology and soil fertility of degraded grasslands at Cherrapunji in North Eastern India," Environment Conservation, vol. 15, pp. 29–35, 1988.

6. M. Kellman, K. Miyanishi, and P. Hiebert, "Nutrient retention by savanna ecosystems II. Retention after fire," Journal of Ecology, vol. 73, no. 3, pp. 953–962, 1985.

7. I. F. Ahlgren and C. E. Ahlgren, "Effect of prescribed burning on soil microorganisms in a Minnesota jack pine forest," Ecology, vol. 46, no. 3, pp. 304–310, 1965.

8. G. Griffith, "Fertility problems in Uganda," Technical Communication, Commonwealth Bureau of Soil Science, vol. 46, pp. 160–164, 1949.

9. M. Ram and B. P. Singh, "Soil fertility management in farming systems," Lectures notes, off campus training on farming system, Aizawl, India, 1993.

10. A. Singh and M. D. Singh, "Effect of various stages of shifting cultivation on soil erosion from steep hill slopes," Indian Forester, vol. 106, no. 2, pp. 115–121, 1981.

11. D. Mandal, K. S. Dadhwal, O. P. S. Khola, and B. L. Dhyani, "Adjusted T values for conservation planning in Northwest Himalayas of India," Journal of Soil and Water Conservation, vol. 61, no. 6, pp. 391–397, 2006.

12. U. C. Sharma, "Methods of selecting suitable land use system with reference to shifting cultivation in NEH region," Indian Journal of Soil Conservation, vol. 26, no. 3, pp. 234–238, 1998.

13. N. P. Singh, O. P. Singh, and N. S. Jamir, Sustainable Agriculture Development Strategy for North Eastern Hill Region of India, Mittal, New Delhi, India, 1996, Edited by Shukla S. P. and Sharma N.

14. B. P. Bhatt, "Agroforestry for sustainable mountain development in N.E.H. region," in Central Himalaya Environment and Development: Potentials, Actions and Challenges, M. S. S. Rawat, Ed., pp. 206–223, Transmedia Publisher, Uttaranchal, India, 2003.

15. Umashankar, "Indigenous agroforestry tree species for conservation and rural livelihood," inAgroforestry in North East India: Opportunities and Challenges, B. P. Bhatt and K. M. Bujarbaruah, Eds., pp. 149–174, ICAR Research Complex for NEH Region, Umiam, Meghalaya, 2005.

16. R. Saha, J. M. S. Tomar, and P. K. Ghosh, "Evaluation and selection of multipurpose tree for improving soil hydro-physical behaviour under hilly eco-system of north east India," Agroforestry Systems, vol. 69, no. 3, pp. 239–247, 2007.

17. B. P. Singh and S. K. Dhyani, "Significance of jackfruit in restoration of soil fertility," Annual Report, ICAR Research Complex, Umiam, Meghalaya, 1995.

18. B. Majumdar, K. Kumar, M. S. Venkatesh, Patiram, and Bhatt B. P., "Effect of different agroforestry systems on soil properties in acid Alfisols of Meghalaya," Journal Hill Research, vol. 17, no. 1, pp. 1–5, 2004.

19. B. Majumdar, M. S. Venkatesh, K. K. Satapathy, K. Kumar, and Patiram, "Effect of alternative farming systems to shifting cultivation on soil fertility," Indian Journal of Agricultural Sciences, vol. 72, no. 2, pp. 122–124, 2002.

20. R. Saha and V. K. Mishra, "Long-term effect of various land use systems on physical properties of silty clay loam soil of N-E hills," Journal of the Indian Society of Soil Science, vol. 55, no. 2, pp. 112–118, 2007.

21. K. K. Satapathy, "Runoff production on hill slopes under different land use systems," inAgroforestry in North East India: Opportunities and Challenges, B. P. Bhatt and K. M. Bujarbaruah, Eds., pp. 451–459, ICAR Research Complex for NEH Region, Umiam, Meghalaya, 2005.

22. R. Saha, V. K. Mishra, and J. M. S. Tomar, "Effect of agroforestry systems on erodibility and hydraulic properties of Alfisols in eastern Himalayan region," Indian Journal of Soil Conservation, vol. 33, pp. 251–253, 2005.

23. R. Saha, P. K. Ghosh, V. K. Mishra, B. Majumdar, and J. M. S. Tomar, "Can agroforestry be a resource conservation tool to maintain soil health in the fragile ecosystem of north-east India?"Outlook on Agriculture, vol. 39, no. 3, pp. 191–196, 2010.

24. M. Datta and N. P. Singh, "Growth characteristics of multipurpose tree species, crop productivity and soil properties in agroforestry systems under subtropical humid climate in India," Journal of Forestry Research, vol. 18, no. 4, pp. 261–270, 2007.

25. S. L. Swamy, S. Puri, and A. K. Singh, "Growth, biomass, carbon storage and nutrient distribution in Gmelina arborea Roxb. Stands on red lateritic soils in central India," Bioresource Technology, vol. 90, no. 2, pp. 109–126, 2003.

26. P. K. Ghosh, A. Das, R. Saha et al., "Conservation agriculture towards achieving food security in North East India," Current Science, vol. 99, no. 7, pp. 915–921, 2010.

27. R. Saha and V. K. Mishra, "Effect of organic residue management on soil hydro-physical characteristics and rice yield in eastern Himalayan region, India," Journal of Sustainable Agriculture, vol. 33, no. 2, pp. 161–176, 2009.

28. V. K. Mishra, R. Saha, and K. M. Bujarbaruah, "Zero tillage technique for transplanted rice in high rainfall eco-system," Scientific leaflet., ICAR Research Complex for NEH Region, Umiam, Meghlaya, 2005.

29. R. Saha, V. K. Mishra, B. Majumdar, K. Laxminarayana, and P. K. Ghosh, "Effect of integrated nutrient management on soil physical properties and crop productivity under a Maize (Zea mays)-mustard (Brassica campestris) cropping sequence in acidic soils of Northeast India,"Communications in Soil Science and Plant Analysis, vol. 41, no. 18, pp. 2187–2200, 2010.

30. P. K. Ghosh, R. Saha, J. J. Gupta et al., "Long-term effect of pastures on soil quality in acid soil of north-east India," Australian Journal of Soil Research, vol. 47, no. 4, pp. 372–379, 2009.

31. K. Laxminarayana, B. P. Bhatt, and R. Tulsi, "Soil fertility buildup through hedgerow intercropping in integrated farming system: a case study," in Agroforestry in North East India: Opportunities and Challenges, B. P. Bhatt and K. M. Bujarbaruah, Eds., pp. 495–506, ICAR Research Complex for NEH Region, Umiam, Meghalaya, 2005.

Chapter 3

Effects of Stubble Management on Soil Fertility and Crop Yield of Rainfed Area in Western Loess Plateau, China

G. B. Huang[1], Z. Z. Luo[2], L. L. Li[1], R. Z. Zhang[2], G. D. Li[3], L. Q. Cai[2], and J. H. Xie[1]

[1]Gansu Provincial Key Laboratory of Arid Land Crop Science/Faculty of Agronomy, Gansu Agricultural University, Lanzhou 730070, China

[2]Faculty of Resource and Environmental Science, Gansu Agricultural University, Lanzhou 730070, China

[3]E H Graham Centre for Agricultural Innovation (Alliance between NSW Department of Primary Industries and Charles Sturt University), Wagga Wagga Agricultural Institute, PMB, Wagga Wagga, NSW 2650, Australia

ABSTRACT

The combination of continuous cereal cropping, tillage and stubble removal reduces soil fertility and increases soil erosion on sloping land. The objective of the present study was to assessment soil fertility changes under stubble removal and stubble retention in the Loess Plateau where soil is prone to severe erosion. It was indicated that soil N increased a lot for and two stubble retention treatments had the higher N balance at the end of two rotations. Soil K balance performed that soil K was in deficient for all treatments and two stubble retention treatments had lower deficit K. The treatments with stubble retention produced higher grain yields than the stubble removal treatments. It was concluded that stubble retention should be conducted to increase crops productivity, improve soil fertility as well as agriculture sustainability in the Loess plateau, China.

INTRODUCTION

Crop stubble is a main agricultural waste material as well as a renewable resource, due to being rich in nitrogen (N), phosphorus (P), and potassium (K). China has a long tradition of efficient recycling of organic residues in agriculture, but this tradition is rapidly disappearing following the intensification of agricultural production, the increased use of mineral fertilizers, and the increasing urbanization and decoupling of crop production and animal production [1]. The intensification of agricultural production has greatly increased the agricultural production, but at the same time, it has contributed to a decrease in resource use efficiency, land degradation through increased wind and water erosion, and pollution of ground water and surface waters [2–4]. There are approximately 0.7 billion ton of organic residues produced each year in China, which contain 3, 0.70, and 7 million ton of N, P, and K, respectively, equivalent to 25% of the total chemical fertilizers used for farming system [5].

In the last few decades, there has been increased interest in the reuse of crop stubble for soil ecology [6, 7], crop system [8, 9], and atmospheric environment [10] worldwide. Retention of plant residues has been found to have many long-term benefits around the world. These crop stubble constitutes a mulch cover that protects the soil against run-off and erosion [11] and increases the percentage of organic matter in the surface soil layer [12,13]. Nutrient loss due to runoff is also decreased [14]. The capacity of the soil surface to intercept rainfall is improved because of changes in soil roughness, soil surface porosity, and hydraulic conductivity of the topsoil. Mulching also reduces temperature extremes [15, 16] and direct evaporation [17, 18]. As a result, crop productivity is often improved. However, according to existing problems of rational and effective utilization of stubble resources under different soils and climatic conditions, the choice of the best suited utilization of stubble must be harmonious to particular agroecological environment.

On the western Loess Plateau in China, dryland cropping systems are dominated by wheat. The practice of 3 ploughs and 2 harrows is employed prior to sowing to prepare a seedbed, while all crop stubble and residues are normally removed from the field at harvest for animal feed or fuel for heating or cooking [19]. The combination of continuous cereal cropping, tillage, and stubble removal reduced soil fertility and increased soil erosion on sloping land [20, 21]. However, little research on stubble retention had been undertaken in the semiarid areas on the western Loess Plateau where soil is prone to severe erosion. The objective of the present study was to assess soil fertility changes under stubble removal and stubble retention in the Loess Plateau.

METHODS AND MATERIALS

Site Description

The field experiment was conducted from 2001 to 2009 at the Dingxi Experimental Station (35°28'N, 104°44'E, elevation 1971m a.s.l.) of Gansu Agricultural University, Anding County, Gansu Province, northwest China. The site had a Huangmian soil [22], aligning with a Calcaric Cambisols in the FAO soil map of the world [23]. It is a sandy loam with low fertility. Soil organic carbon was below 7.63gk (Table 1), representing the major cropping soil in the district [24], one of two dominant soils on the Loess Plateau. Long-term annual rainfall at Dingxi averages 391mm, ranging from 246mm in 1986 to 564mm in 2003, with about 54% received between July and September. Daily maximum temperatures can reach up to 38°C in July, while minimum temperatures can drop to −22°C in January. Hence, summers are warm and moist, whereas winters are cold and dry. Annual accumulated temperature >10°C is 2239°C, and annual radiation is 5929MJm^{-2}

with 2477 h of sunshine. The site had a long history of continuous cropping using conventional tillage. The crop prior to the experiment commencement in 2001 was flax (Linum usitatissimum L.).

Table 1: Soil chemical and physical properties at the start of experiment

Depth (cm)	Bulk density (g cm−3)	Organic matter (g kg−1)	Total N (g kg−1)	Total P (g kg−1)	Olsen P (mg kg−1)	Available K (mg kg−1)	pH
0–5	1.29	13.15	0.85	0.83	5.81	290.09	8.30
5–10	1.23	12.86	0.87	0.84	5.02	274.00	8.40
10–30	1.32	11.95	0.78	0.79	2.14	202.47	8.30

Experimental Design and Treatment Description

The experiment had a fully phased factorial design with 2 phases, replicated 4 times (blocks). Spring wheat (cv. Dingxi no. 35) and field pea (cv. Yannong) were sown in rotation in both phases represented in each year. Phase 1 (P/W) started with field pea followed by spring wheat, and phase 2 (W/P) started with spring wheat followed by field pea. Therefore, there were 32 plots in total. Plots were 4 m wide, 17 m long in block 1, 21 m long in blocks 2 and 3, and 20 m long in block 4. All treatments were described as follows (Table 2).

Table 2: Details of treatments used in the long-term conservation tillage experiment

Code	Treatments	Description
T	Conventional tillage with stubble removed	Fields were ploughed 3 times and harrowed twice after harvesting. The first ploughing was in August immediately after harvesting, the second and third ploughing were in late August and September, respectively. The plough depths were 20 cm, 10 cm, and 5 cm, respectively. The field was harrowed after the last cultivation in September and again in October before the ground was frozen. This is the typical conventional tillage practice in the Dingxi region.
NT	No-till with stubble removed	No-till throughout the life of the experiment. The straw was removed from the field and used as fuel or feed.
TS	Conventional tillage with stubble incorporated	Fields were ploughed and harrowed exactly as for the T treatment (3 passes of plough and 2 harrows), but with straw incorporated at the first ploughing. All the straw from the previous crop was returned to the original plot immediately after threshing and then incorporated into the ground.

NTS	No-till with stubble cover	No-till throughout the life of the experiment. The ground was covered with the straw of previous crop from August until the following March. All the straw from previous crop was returned to the original plot immediately after threshing.

Sowing Rate, Fertilizers, and Field Management

All crops were sown by a small no-till seeder (5-6 rows in 1.2 m width) designed by the China Agricultural University. The no-till seeder, drawn by a 13.4 kW (18 HP) tractor, was designed to place fertilizers below the seeds using narrow points followed by concave rubber press wheels in one operation. Spring wheat was sown at 187.5 kg ha^{-1} in mid-March and harvested in late July to early August each year. Field pea was sown at 180 kg ha^{-1} in early April and harvested in early July each year. The row spacing was 20 cm for spring wheat and 24 cm for field pea using the no-till seeder.

Nitrogen and P were applied at 105 kg N ha^{-1} as urea (46% N) and at 45.9 kg P ha^{-1} as calcium superphosphate (6.1% P) for spring wheat, and 20 kg N ha^{-1} and 45.9 kg P ha^{-1} for field pea. No farm manure was used in this experiment. Field peas were not inoculated when sown as no appropriate rhizobia were available on the market. However, the site had history of field pea in the previous 3 years. Roundup (glyphosate, 10%) was used for weed control during fallow after harvesting as per the product guidelines. During the growing season, weeds were removed by hand. Pests and diseases were monitored and controlled as per conventional practice in the area.

Measurements

Soil Properties

Soil samples were collected from 0–30 cm depth for the determination of soil nutrient level after harvest (mid-August, 2007). Five cores were bulked into one sample for each plot using a standard 25 mm diameter soil corer. Soil sample were dried and sieved through 2 mm mesh. Soil organic carbon was determined by dichromate oxidation [25]. Total N in soil was determined by Semimicro-Kjeldahl method [26]. Nitrate nitrogen (NO_3^--N) and exchangeable ammonium nitrogen (NH_4^+-N) in soil was determined using $FeSO_4/Zn$ reduction method described by Carter [27]. Total P in soil was determined using Sodium carbonate fusion described by Carter [27]. Available phosphorus (P) in soil was determined by extracting samples with $0.5 M NaHCO_3$, and determining P colorimetrically using molybdate [28]. Total K in soil was determined using Flame photometry method [28]. Available K in soil was determined using 1 N ammonium acetate extraction-flame photometry method [28]. Nitrogen in grain and crop residues (straw and chaff) was determined using the method described by Lu [29]. Potassium in grain and crop residues (straw and chaff) was determined using the method described by Bao [28].\

Nitrogen Fixation

Nitrogen fixation by field pea was estimated in 2005 using the method of ^{15}N natural abundance as described by Armstrong et al. [30]. At anthesis, 5 individual field pea plants were cut at ground level from each plot, bulked into one sample and dried at 60°C for 24 h. At the same time, 5 nonlegume plants (weeds) from the plot were also collected and oven-dried at 60°C as "reference plants". Both the legumes and reference plants were ground through 1 mm mesh, then subsampled and finely ground prior to analysis of ^{15}N natural abundance using continuous flow isotope ratio mass spectrometry (Europa Scientific ANCA System) [31].

Grain Yield

The whole plot was harvested manually using sickles at 5 cm above ground. The edges (0.5 m) of the plot were trimmed and discarded. Samples were then processed to obtain grain yield, straw and chaff. The grain yield from the harvesting area was recorded and converted to yield per hectare.

Calculations

Nitrogen Balance

Nitrogen balance was calculated over 4 years with two complete rotation cycles. Nitrogen inputs included N in fertilizers and N in seeds. The N in straw brought into the system (6.8 t/ha) in 2002 was also taken into account for TS and NTS treatments. Nitrogen output includes grain N and stubble N if stubble was removed (e.g., T and NT treatments). Nitrogen fixed by field pea in 2001–2004 was extrapolated using data in 2005 as no data were available in 2001–2004.

Nitrogen Fixation

Nitrogen fixation by field pea was calculated as follows:

$$\%Ndfa = \frac{\left(\delta^{15}N_{weeds} - \delta^{15}N_{legume}\right)}{(\delta^{15}N_{weeds} - \beta)}, \tag{1}$$

(see [30]), where %Ndfa is the percentage of plant total N derived from fixation, $\delta^{15}N_{weeds}$ is the natural abundance of ^{15}N in reference plant (weeds), $\delta^{15}N_{legume}$ is the natural abundance of ^{15}N in legume (field pea), and β represents a measure of the isotopic fraction associated with redistribution of N between roots and shoots.

Potassium Balance

Potassium balance was calculated after 5 years. Potassium inputs included K in straw and K in seeds. Potassium output includes grain K and stubble K if stubble was removed (e.g., T and NT).

Data Analysis

Analysis of variance was performed to determine the effects of different stubble management on soil fertility and grain yield. All statistical analyses of data were carried out through the SPSS package.

RESULTS AND DISCUSSION

The Distribution of Soil Organic Matter and Total Nutrients under Different Stubble Management

Results showed that stubble retention increased soil organic matter at the 0–5 cm and 10–30 cm depth significantly (P < 0.01), while increased soil organic matter at the 5–10 cm significantly (P < 0.001) after 6 years in W/P rotation sequence (Table 3). In the top 5 cm depth, soil organic matter under NTS was the highest, while soil organic matter under TS was the highest at lower depths. The pattern for organic matter distribution in the W/P rotation sequence was similar to that in the P/W rotation sequence. Soil organic matter concentration was relatively uniformly distributed within the 0–30 cm depth under TS treatment. In contrast, NTS treatment resulted in a significant increase in soil organic matter at the soil surface.

Table 3: Soil total nutrients as affected by stubble management in different rotation sequence (g kg⁻¹)

Depth (cm)	Treatment	Field pea-spring wheat (P/W)				Spring wheat-field pea (W/P)			
		Organic matter	Total N	Total P	Total K	Organic matter	Total N	Total P	Total K
0–5	T	13.90 ± 0.64	1.00 ± 0.05	0.85 ± 0.03	15.08 ± 0.64	13.30 ± 0.52	1.00 ± 0.04	0.90 ± 0.07	15.58 ± 0.76
	NT	14.01 ± 0.79	1.05 ± 0.02	0.90 ± 0.15	14.61 ± 0.34	13.78 ± 1.17	1.04 ± 0.03	0.96 ± 0.04	14.48 ± 0.34
	TS	15.26 ± 0.43	1.09 ± 0.03	1.06 ± 0.12	16.31 ± 0.92	14.56 ± 0.62	1.09 ± 0.04	1.07 ± 0.02	16.36 ± 0.33
	NTS	15.66 ± 1.12	1.13 ± 0.02	1.08 ± 0.04	16.77 ± 0.35	16.81 ± 1.50	1.21 ± 0.11	1.09 ± 0.03	16.92 ± 0.14
	Significant	**	*	*	*	***	*	**	**

5–10								
T	13.83 ± 0.68	0.99 ± 0.04	0.80 ± 0.02	15.08 ± 0.60	13.14 ± 0.32	0.99 ± 0.03	0.85 ± 0.06	14.47 ± 0.71
NT	14.00 ± 0.58	1.04 ± 0.04	0.82 ± 0.02	14.13 ± 0.22	13.70 ± 0.87	1.01 ± 0.02	0.95 ± 0.14	13.98 ± 0.61
TS	15.11 ± 0.38	1.04 ± 0.04	1.00 ± 0.09	15.47 ± 1.73	14.24 ± 0.84	1.01 ± 0.01	0.99 ± 0.02	15.67 ± 0.26
NTS	14.78 ± 0.17	1.07 ± 0.01	1.02 ± 0.13	16.17 ± 0.33	14.94 ± 0.62	1.12 ± 0.09	1.03 ± 0.08	16.31 ± 0.26
Significant	***	ns	*	ns	***	*	ns	**

10–30								
T	13.17 ± 0.79	0.99 ± 0.04	0.81 ± 0.06	14.53 ± 0.40	12.42 ± 1.09	0.96 ± 0.06	0.871 ± 0.03	14.54 ± 0.48
NT	13.70 ± 0.82	0.99 ± 0.03	0.81 ± 0.05	13.86 ± 0.43	13.49 ± 0.48	1.01 ± 0.05	0.87 ± 0.12	14.21 ± 0.37
TS	14.45 ± 0.59	1.00 ± 0.03	0.83 ± 0.04	15.10 ± 1.05	14.03 ± 0.91	1.01 ± 0.03	0.89 ± 0.02	15.27 ± 0.85
NTS	14.15 ± 0.26	1.00 ± 0.05	0.84 ± 0.02	15.45 ± 0.66	13.80 ± 0.75	1.02 ± 0.04	0.86 ± 0.02	15.49 ± 0.04
Significant	**	ns	ns	ns	*	ns	ns	ns

Significant level (ns: not significant; *$P < 0.05$; **$P < 0.01$; ***$P < 0.001$).

Accumulation of soil organic matter at the soil surface was a result of surface placement of crop residues and a lack of soil disturbance that kept residues isolated from the rest of the soil profile. Decomposition of surface-placed residues is often slower than when incorporated in the soil profile [32, 33], primarily because of less optimal moisture conditions [34]. The apparent soil organic matter accumulation in stubble retention treatments noted by these results is consistent with the findings of Lao et al. [35] and Lin et al. [36]. Sun et al. [7] reported that soil organic matter increased 15.8%~18.1%, 6.6%~10.6%, and 1.3%~1.9% stubble retention compared with straw removal in the 0–10, 10–20, 20–40 cm depths in Wushan soil after 15 years.

Stubble-induced changes in soil total N are often directly related to changes in soil organic C. This similarity may be related to soil organic matter which could influence nutrient retention and supply [37]. Both stubble retention and no-tillage increased soil total N concentration significantly ($P < 0.05$) at the soil surface 0–5 cm (Table 3), but it was uniformly distributed with depths under T treatment. In the 5–10 cm, both stubble retention and no-tillage increased soil total N concentration significantly ($P < 0.05$) in W/P rotation sequence, but not in P/W rotation sequence. In the 10–30 cm depths, there was no difference in soil total N concentration ($P > 0.05$) among stubble management and tillage systems in either rotation sequences. Soil total N probably contains compounds that are more resistant to decomposition and which consequently can affect N dynamics in the soil. Reeves et al. [38] reported that under stubble management and no-tillage system, soil N losses were reduced, but short-term N availability was also reduced. The higher total N content associated with stubble retention treatments in this study can also be attributed to a reduction in soil erosion. The amount of soil lost due to soil erosion has been reported as high as 3720 t km^{-2} year^{-1}, rising to maxim of 3720 t km^{-2} year^{-1}, in this area during rainy season (July–September) [21]. The stubble retention treatments leaves crop residues on the soil surface and creates more large size soil aggregates, thus reducing soil erosion and contributing to the higher organic carbon and total N content in soil. In addition, because stubble yield was different for TS and NTS, the amount of crop stubble returned to the soil should be different as well. Thus, difference in total N was not only due to difference in disturbance by tillage operations, but also due to difference in the amount of crop residue returned to the soil under TS and NTS.

Unlike soil C and N, soil P does not readily undergo oxidation-reduction reactions in the common processes of organic matter decomposition. Total P concentration was significantly ($P < 0.05$) greater in the top 5 cm depth of stubble retention treatments in P/W rotation sequence, while significantly ($P < 0.01$) greater in the top 5 cm depth of stubble retention treatments in W/P rotation (Table 3). In the 5–10 cm depth, total P concentration was significantly ($P < 0.05$) greater in P/W rotation sequence, but not in W/P rotation sequence. In the 10–30 cm depths, there was no difference in soil total P concentration ($P > 0.05$) among residue management and tillage systems in either rotation sequences. NTS had the lowest total P concentration in W/P rotation sequence. Surface application of phosphate fertilizer, immobilization of phosphate fertilizer as well as stubble cover in surface soil may account for this result. It has been proposed that organic matter itself may be considered an important source for P recycling in the short and long term [39].

Total K was significantly ($P < 0.05$) greater in the top 5 cm depth of stubble retention treatments in P/W rotation sequence, while significantly ($P < 0.01$) greater in the top 5 cm depth of stubble retention treatments in W/P rotation (Table 3). In the second 5 cm, total K concentration was only significantly ($P < 0.01$) greater in W/P rotation sequence, but not in P/W rotation sequence. In the 10–30 cm depth, although NTS and TS showed higher total K concentration, there was no difference ($P > 0.05$) among stubble management and tillage systems in either rotation sequences. This result is due to output of soil K removed out of cropland with crop straw, which could be returned back to soil by organic material recycling in stubble retention treatments. It has been proposed that straw itself may be considered an important K source, which could return back 70~80% of soil K removed from cropland with crop straw [9].

The Distribution of Soil Available Nutrients under Different Stubble Management

Results showed that soil available N concentration was higher in the top 5 cm depth of stubble retention, while higher in the 10–30 cm depth of stubble removal treatment among residue management and tillage systems in either rotation sequences. However, there were no difference in soil available N concentration (P > 0.05) among stubble management and tillage systems, except for that in the top 5 cm of W/P rotation sequence and 10–30 cm of P/W rotation sequence (Table 4). In this experiment, available N referred to nitrate-N and ammonia-N. Residue retention could reduce gaseous loss by volatilization of ammonia-N arisen from soil alkalinity and increased temperatures [40, 41], while denitrification and leach of nitrate-N, on the other hand, could be very serious with a combination of no tillage due to improved soil moisture [42–45].

Table 4: Soil available nutrients as affected by stubble management in different rotation sequence (mg kg^{-1})

Depth (cm)	Treatment	P/W			W/P		
		Available N	Available P	Available K	Available N	Available P	Available K
0–5	T	37.37 ± 1.77	13.61 ± 0.84	211.41 ± 21.28	37.31 ± 0.94	14.22 ± 0.82	192.04 ± 13.02
	NT	33.94 ± 1.97	13.78 ± 0.55	247.13 ± 12.49	35.63 ± 0.90	14.25 ± 2.50	227.91 ± 21.18
	TS	37.97 ± 0.92	16.40 ± 1.34	263.76 ± 38.74	40.64 ± 1.82	16.23 ± 0.76	265.17 ± 16.01
	NTS	37.44 ± 1.02	18.96 ± 1.46	280.66 ± 47.11	40.76 ± 1.86	18.68 ± 2.29	286.98 ± 41.57
	Significant	ns	***	ns	*	*	*

5–10	T	35.13 ± 3.02	12.34 ± 1.10	190.74 ± 19.19	35.94 ± 2.25	14.95 ± 1.02	187.41 ± 37.75
	NT	34.54 ± 0.99	13.12 ± 1.45	202.43 ± 32.17	34.51 ± 5.88	15.09 ± 0.89	191.90 ± 26.75
	TS	36.77 ± 1.84	14.15 ± 1.38	259.75 ± 4.02	39.63 ± 0.98	16.30 ± 2.22	248.04 ± 34.31
	NTS	34.61 ± 2.03	16.04 ± 2.12	262.65 ± 24.37	37.95 ± 2.65	16.11 ± 0.58	265.27 ± 45.85
	Significant	ns	*	*	ns	ns	ns
10–30	T	41.91 ± 0.09	5.57 ± 2.34	159.45 ± 9.27	42.44 ± 2.61	5.15 ± 0.65	139.25 ± 5.39
	NT	37.43 ± 1.95	4.76 ± 0.72	163.82 ± 6.59	37.99 ± 1.00	4.74 ± 0.56	147.90 ± 16.71
	TS	41.38 ± 0.98	5.74 ± 0.39	195.07 ± 9.71	42.42 ± 2.46	5.24 ± 0.38	192.27 ± 6.81
	NTS	37.91 ± 2.50	5.57 ± 0.44	202.21 ± 20.53	40.77 ± 2.48	4.96 ± 0.61	202.03 ± 46.50
	Significant	*	ns	*	ns	ns	*

Significant level (ns: not significant; $*P < 0.05$; $**P < 0.01$; $***P < 0.001$).

Available P concentration was significantly ($P < 0.001$) greater in the top 5 cm depth of stubble retention treatments in P/W rotation sequence, while significantly ($P < 0.05$) greater in the top 5 cm depth of stubble retention treatments in W/P rotation (Table 4). In the 5–10 cm depth, available P concentration was significantly ($P < 0.05$) greater in P/W rotation sequence, but not in W/P rotation sequence. In the 10–30 cm depths, there was no difference in soil available P concentration ($P > 0.05$) among residue management and tillage systems in either rotation sequences. Surface application of phosphate fertilizer, immobilization of phosphate fertilizer as well as stubble cover in surface soil may account for this result. P solubility is known to be enhanced by increasing SOM and decreasing pH in alkaline soils [46] by acidifying the rhizosphere soil. Where crop residues are returned to the soil, an increase in P availability may occur by decreasing the adsorption of P to mineral surfaces [47] which complements biologically mediated release of P to improve crop P status. With time, soil and crop residue management practices that promote organic matter accumulation would be expected to improve P nutrition of crops.

Available K concentration was only significantly ($P < 0.05$) greater in the top 5 cm depth of stubble retention treatments in W/P rotation sequence, but not in P/W rotation sequence (Table 4). In the 5–10 cm, available K concentration was only significantly ($P < 0.05$) greater in rotation P/W sequence, but not in W/P rotation sequence. In the 10–30 cm depth, there was significantly ($P < 0.05$) difference among stubble management in either rotation sequences. Intermediate or final products involving organic acids and CO_2 produced by SOM could improve availability of fixed K. Where crop residues are returned to the soil, an increase in K availability may occur by decreasing the adsorption of K to clay mineral surfaces [43]. In this case, quality of available nutrients for seasoning crop could be enhanced by residue management and tillage systems, thus ensuring improvement of grain yields [48].

Soil Nutrient Balance under Different Stubble Management

TS and NTS had higher total N input than T and NT due to extra N from previous straw (Table 5), but no difference between TS and NTS and between T and NT in either rotation sequences. However, the total N output was significantly ($P < 0.01$) different between treatments in either rotation sequences. The treatments with high harvest dry matter and higher grain yield had more N% output from the system. However, the most of N could be returned back to soil by organic material recycling in stubble retention treatments. The T treatment exported the greatest N from crop harvest, whereas TS treatment had the lowest over 4 years (Table5). N fixation by field pea under different treatments was also significantly ($P < 0.001$) different in either rotation sequences (Table 5). As a result, over 4 years all treatments accumulated N in soils, and there was also significant ($P < 0.001$) difference under different treatments in either rotation sequences (Table 5). Treatments, retained the crop stubble, had high N balance at the end of two rotation cycles. Therefore, stubble retention could improve soil N balance year by year. But as a whole, soil N increased from 2002 to 2005 for both phases, indicating total N fertilizer applied in the field were more than crop needed as evidenced by high N balance at the end of two rotation cycles. Therefore, it is recommended that stubble retention should be practiced to increase crops productivity and improve soil N storage and N fertilizer use should be reduced. High N input not only increased input costs, but also increased the risk of environmental contamination in the ground water system. In the current research, there was up to 185 kg ha^{-1} surplus N (i.e., W/P rotation sequence) accumulated after two rotation cycles under the current fertilizer regime over the 4 years. It appears to be excellent prospects for reducing current farmer fertilizer N inputs while maintaining spring wheat yields in all seasons. This is a significant finding as it will directly increase farm profitability with little risk. Considerable savings are likely by adopting optimum fertilizer N rates. There appear to be no other reports on estimates of N fixation for the Loess Plateau.

Table 5: Nitrogen balance under different treatments over 4 years (kg N ha^{-1})

Rotation	Item	T	NT	TS	NTS	Significant
P/W	Input	269.00	269.00	299.37	299.37	—
	Fixation	4.65	27.12	20.73	26.06	***
	Output	183.45	167.18	139.73	159.25	**
	Balance	90.20	128.94	180.37	166.18	***
W/P	Input	269.00	269.00	299.37	299.37	—
	Fixation	3.42	25.63	16.90	24.89	***
	Output	144.82	130.34	102.51	139.09	**
	Balance	127.60	164.29	213.76	185.17	***

Significant level (*P < 0.05; **P < 0.01; ***P < 0.001).

No K fertilizer and farm manure was applied in this experiment. In P/W rotation sequence, total K input was same for T and NT, but more K input for TS and NTS due to K input from previous straw. However, total K output was significantly (P < 0.01) different between treatments. The higher harvest dry matter and higher grain yield, the higher K output from the system. NTS treatment exported the greatest K from crop harvest, whereas NT treatment had the lowest K. As a result, there was also significant (P < 0.01) different in soil K balance under different treatments (Table 6). Soil K balance in W/P rotation sequence was similar to that in P/W rotation sequence. However, the average K balance under T treatment in W/P rotation sequence was higher than that in P/W rotation sequence, whereas the average K balance under TS, NT, and NTS treatments in W/P rotation sequence was lower than that in P/W rotation sequence. Unlike soil N balance being in surplus, soil K was indeficient for all treatments, ranged from 11.25 kg K ha^{-1} for the NTS treatment to 36.95 kg K ha^{-1} for the T treatment (Table 6). Therefore, it is suggested that fertilizer K should be applied to maintain soil K balance.

Table 6: Potassium balance under different treatments over 5 years (kg K ha^{-1})

Rotation	Item	T	NT	TS	NTS	Significant
P/W	Input	4.93	4.65	40.15	49.27	—
	Output	41.88	33.01	52.48	60.52	**
	Balance	−36.95	−28.36	−12.32	−11.25	**

	Input	0.68	0.68	14.68	19.03	—
W/P	Output	27.82	31.38	35.99	39.87	**
	Balance	−27.14	−30.70	−21.31	−20.84	**

Significant level(*P < 0.05; **P < 0.01; ***P < 0.001).

Crop Yield under Different Stubble Management

For spring wheat, the treatments with stubble retention produced more grain yields than the treatments with stubble removed in all 8 years. There were significant differences in grain yield among treatments, except that in 2005, 2006, and 2008 (Table 7). For field pea, there were significant differences among treatments, except that in 2004, 2005, 2008, and 2009. Averaged across 8 years, yield of wheat under the NTS and TS treatment was 21% (0.344 t ha^{-1}) and 9% (0.144 t ha^{-1}) higher than the T treatment. Similarly, yield of field pea under the NTS and TS treatment was 20% (0.227 t ha^{-1}) and 2% (0.026 t ha^{-1}) higher than the T treatment. A recent survey of farmers in the area surrounding the experimental site found average grain yield was just 1.0 t ha^{-1} for spring wheat and 0.8 t ha^{-1} for field pea in the 2003 season [49]. Corresponding results from the current study for T and NTS were 1.641 t ha^{-1} and 1.986 t ha^{-1} for spring wheat, and 1.125 t ha^{-1} and 1.351 t ha^{-1} for field pea, respectively. This represents a potential significant improvement in grain yield of 40%~69% for field pea and 64%~99% for spring wheat through better agronomic practices (e.g., stubble retention). These results were inconsistent with the findings of McCalla and Army [50] who reported that stubble retention decreased yields, especially in a humid climate, due to poor crop establishment under reduced tillage [51]. Graham et al. [52] found that in autumn-sown crops under direct drill, the yields were lower when straw residues were left on the surface than when residues were burnt. Incorporation of straw reduced the detrimental effect of straw on yields, but the yields were lower than those where the straw had been burnt.

Table 7: Grain yield under different treatments (tha^{-1})

Crop	Year	T	NT	TS	NTS	Significant
Spring wheat	2002	1.816 ± 0.279	1.414 ± 0.362	1.736 ± 0.276	2.151 ± 0.246	***
	2003	1.416 ± 0.281	1.545 ± 0.356	1.646 ± 0.367	1.825 ± 0.132	*
	2004	2.189 ± 0.248	1.664 ± 0.219	2.162 ± 0.221	2.382 ± 0.304	*
	2005	2.900 ± 0.519	3.077 ± 0.292	2.988 ± 0.663	3.327 ± 0.060	ns
	2006	1.383 ± 0.210	1.317 ± 0.200	1.565 ± 0.235	1.549 ± 0.123	ns
	2007	0.562 ± 0.132	0.633 ± 0.169	0.666 ± 0.126	0.944 ± 0.187	**
	2008	1.632 ± 0.549	1.818 ± 0.899	1.851 ± 0.312	2.100 ± 0.329	ns
	2009	1.233 ± 0.371	0.985 ± 0.644	1.670 ± 0.325	1.607 ± 0.383	*

Field	2002	1.653 ± 0.177	1.416 ± 0.275	1.527 ± 0.313	1.790 ± 0.213	*
pea	2003	0.881 ± 0.206	0.803 ± 0.156	0.823 ± 0.101	1.269 ± 0.288	**
	2004	1.708 ± 0.145	1.496 ± 0.440	1.681 ± 0.349	1.668 ± 0.193	ns
	2005	1.686 ± 0.24	1.816 ± 0.268	1.911 ± 0.672	2.119 ± 0.534	ns
	2006	0.759 ± 0.129	0.552 ± 0.122	0.872 ± 0.123	0.890 ± 0.048	**
	2007	0.206 ± 0.023	0.277 ± 0.067	0.342 ± 0.053	0.553 ± 0.088	***
	2008	1.342 ± 0.196	1.306 ± 0.387	1.190 ± 0.486	1.649 ± 0.180	ns
	2009	0.762 ± 0.127	0.727 ± 0.087	0.857 ± 0.143	0.873 ± 0.249	ns

Significant level l(ns: not significant; *P < 0.05; **P < 0.01; ***P < 0.001).

CONCLUSIONS

Stubble retention on an alkaline soil in a semiarid areas for several years markedly improved fertility and increased crop productivity. Stubble retention resulted in significantly greater soil organic matter at the 0–30 cm depth. Stubble retention increased soil total N significantly at the soil surface 0–5 cm, but soil total N concentration remained similar at 10–30 cm for all treatments. Total P and total K were significantly greater in the 0–10 cm depth under stubble retention treatments. There were no difference in soil available N concentration among stubble management and tillage systems, except that in the top 5 cm of W/P rotation sequence and 10–30 cm of P/W rotation sequence. Available P concentration was significantly greater in the top 5 cm and

5–10 cm depth. Available K concentration was significantly greater at the 0–30 cm depth of stubble retention treatments. Soil N increased greatly from 2002 to 2005 for all treatment. The two stubble retention treatments had the higher N balance at the end of two rotations. Soil K was indeficient for all treatments with more deficit under two stubble retention treatments. As a result, the treatments with stubble retention produced more grain yields than the treatment with stubble removed in all 8 years. Grain yields were the highest under NTS, but the lowest under NT for both spring wheat (1.986 versus 1.557 t ha⁻¹) and field pea (1.351 versus 1.049 t ha⁻¹). It was concluded that stubble retention should be practiced to increase crops productivity, improve soil fertility as well as agriculture sustainability in the Loess plateau.

ACKNOWLEDGMENTS

The project was supported by the Australian Centre for International Agricultural Research (no. CIM-1999-094), the Ministry of Science and Technology in China (no. 2006BAD15B06) and National Natural Science Foundation of China (no. 31060178, and no. 40771132). The authors are grateful to land owners, Mr. J. Q. Li, Mr. F. Zhang, Mr. C. L. Guo, and Mr. C. Q. Wu, for allowing them to use their land and their cooperation, and to Mr. Z. Jin who looked after the site. Authors thank the members of their team for help in experiment. Special thanks to all postgraduate and undergraduate students from Gansu Agricultural University who were involved in this project for different periods.

REFERENCES

1.   X. Ju, F. Zhang, X. Bao, V. Römheld, and M. Roelcke, "Utilization and management of organic wastes in Chinese agriculture: past, present and perspectives," Science in China. Series C, Life sciences / Chinese Academy of Sciences., vol. 48, pp. 965–979, 2005.

2.   F. S. Zhang, W. Q. Ma, W. F. Zhang, and M. Fan, "Nutrient management in China: from production system to food chain," in Plant Nutrition for Food Security, C. J. Li, F. S. Zhang, A. Doberman, et al., Eds., pp. 13–15, Human Health and Environmental Protection, Beijing, China, 2005.

3.   J. Liu and J. Diamond, "China's environment in a globalizing world," Nature, vol. 435, no. 7046, pp. 1179–1186, 2005.

4.   S. Rozelle, J. Huang, and L. Zhang, "Poverty, population and environmental degradation in China,"Food Policy, vol. 22, no. 3, pp. 229–251, 1997.

5.   J. Wang, D. J. Wang, G. Zhang, and C. Wang, "Effects of different nitrogen fertilizer rate with continuous full amount of straw incorporated on paddy soil nutrients," Journal of Soil and Water Conservation, vol. 24, no. 5, pp. 40–44, 2010.

6.   J. A. Ocio, P. C. Brookes, and D. S. Jenkinson, "Field incorporation of straw and its effects on soil microbial biomass and soil inorganic N," Soil Biology and Biochemistry, vol. 23, no. 2, pp. 171–176, 1991.

7.   X. Sun, Q. Liu, D. J. Wang, and B. Zhang, "Effect of long-term application of straw on soil," Fertility Chinese Journal of Eco-Agriculture, vol. 16, no. 3, pp. 587–592, 2008.

8.   A. Whitbread, G. Blair, Y. Konboon, R. Lefroy, and K. Naklang, "Managing crop residues, fertilizers and leaf litters to improve soil C, nutrient balances, and the grain yield of rice and wheat cropping systems in Thailand and Australia," Agriculture, Ecosystems and Environment, vol. 100, no. 2-3, pp. 251–263, 2003.

9.   T. De-shui, J. Ji-yun, H. Shao-wen, L. Shu-tian, and H. Ping, "Effect of Long-Term Application of K Fertilizer and Wheat Straw to Soil on Crop Yield and Soil K Under Different Planting Systems,"Agricultural Sciences in China, vol. 6, no. 2, pp. 200–207, 2007.

10.   H. K. Kludze and R. D. DeLaune, "Straw application effects on methane and oxygen exchange and growth in rice," Soil Science Society of America Journal, vol. 59, no. 3, pp. 824–830, 1995.

11.   D. M. Freebairn and W. C. Boughton, "Hydrologic effects of crop residue management practices,"Australian Journal of Soil Research, vol. 23, no. 1, pp. 23–35, 1985.

12.   A. Roldán, F. Caravaca, M. T. Hernández et al., "No-tillage, crop residue additions, and legume cover cropping effects on soil quality characteristics under maize in Patzcuaro watershed (Mexico)," Soil and Tillage Research, vol. 72, no. 1, pp. 65–73, 2003.

13. K. Y. Chan and D. P. Heenan, "The effects of stubble burning and tillage on soil carbon sequestration and crop productivity in Southeastern Australia," Soil Use Manage, vol. 21, pp. 427–431, 2005.

14. J. R. Smart and J. M. Bradford, "Conservation tillage corn production for a semiarid, subtropical environment," Agronomy Journal, vol. 91, no. 1, pp. 116–121, 1999.

15. B. J. Radford, A. J. Key, L. N. Robertson, and G. A. Thomas, "Conservation tillage increases soil water storage, soil animal populations, grain yield, and response to fertilisers in the semi-arid subtropics,"Australian Journal of Experimental Agriculture, vol. 35, no. 2, pp. 223–232, 1995.

16. K. J. Shinners, W. S. Nelson, and R. Wang, "Effects of residue-free band width on soil temperature and water content," Transactions of the American Society of Agricultural Engineers, vol. 37, no. 1, pp. 39–49, 1994.

17. Z. Y. Liu, C. P. Wang, H. Y. Lu, L. F. Lei, and Y .S. Wu, "The effect of different mulching methods with whole maize straw in dryland on maize yield," Shanxi Agricultural Science, vol. 28, no. 3, pp. 20–22, 2000.

18. J. L. Steiner, "Tillage and surface residue effects on evaporation from soils," Soil Science Society of America Journal, vol. 53, no. 3, pp. 911–916, 1989.

19. R. Lal, "Constraints to adopting no-till farming in developing countries," Soil and Tillage Research, vol. 94, no. 1, pp. 1–3, 2007.

20. T. Fan, B. A. Stewart, W. Yong, L. Junjie, and Z. Guangye, "Long-term fertilization effects on grain yield, water-use efficiency and soil fertility in the dryland of Loess Plateau in China," Agriculture, Ecosystems and Environment, vol. 106, no. 4, pp. 313–329, 2005.

21. G. Liu, "Soil conservation and sustainable agriculture on the Loess Plateau: challenges and prospects,"Ambio, vol. 28, no. 8, pp. 663–668, 1999.

22. Chinese Soil Taxonomy Cooperative Research Group, Chinese Soil Taxonomy (Revised Proposal), Institute of Soil Science/ Chinese Agricultural, Science and Technology Press, Academic Sinica, Beijing, China, 1995.

23. FAO, Soil map of the world: revised legend. World Soil Resources Report 60, Food and Agriculture Organization of the United Nations, Rome, Italy, 1990.

24. Z. Xianmo, L. Yushan, P. Xianglin, and Z. Shuguang, "Soils of the loess region in China," Geoderma, vol. 29, no. 3, pp. 237–255, 1983.

25. A. Walkley and I. A. Black, "An examination of the Degtjareff method for determining soil organic matter, and a proposed modification of the chromic acid titration method," Soil Science, vol. 37, pp. 29–38, 1934.

26. J. M. Bremmer and C. S. Mulvaney, "Total nitrogen," in Methods of Soil Analysis—Part 2. Chemical and Microbiological Properties, C. A. Bluck, Ed., pp. 595–624, American Society of Agronomy, Wis, USA, 1982.

27. M. R. Carter, Soil Sampling and Methods of Analysis, Lewis Publishers, Boca Raton, Fla, USA, 1993.

28. S. D. Bao, Analysis on soil agricultural chemistry, China Agricultural Press, Beijing, China, 2000.

29. R. K. Lu, Methods of Analysis of Soil and Agro-chemistry, China Agricultural Science and Technology Press, Beijing, China, 2000.

30. E. L. Armstrong, J. S. Pate, and M. J. Unkovich, "Nitrogen balance of field pea crops in south western Australia, studied using the 15N natural abundance technique," Australian Journal of Plant Physiology, vol. 21, no. 4, pp. 533–549, 1994.

31. T. Dawson and Brooks P. D., "Fundamentals of stable isotope chemistry and measurement," inApplication of Stable Isotope Techniques to Study Biological Processes and Functioning of Ecosystems, M. J. Unkovich, J. S. Pate, A. M. McNeill, and D. J. Gibbs, Eds., Kluwer Academic, Dodrecht, The Netherlands, 2001.

32. P. L. Brown and D. D. Dickey, "Losses of wheat straw residue under simulated field conditions," Soil Science Society of America Proceedings, vol. 34, pp. 118–121, 1970.

33. F. Ghidey and E. E. Alberts, "Residue type and placement effects on decomposition: field study and model evaluation," Transactions of the American Society of Agricultural Engineers, vol. 36, no. 6, pp. 1611–1617, 1993.

34. A. J. Franzluebbers, M. A. Arshad, and J. A. Ripmeester, "Alterations in canola residue composition during decomposition," Soil Biology and Biochemistry, vol. 28, no. 10-11, pp. 1289–1295, 1996.

35. X. R. Lao, W. H. Sun, and Z. Wang, "Effect of matching use of straw and chemical fertilizer on soil fertility," Acta Pedologica Sinica, vol. 40, no. 4, pp. 618–623, 2033.

36. B. Lin, J. X. Lin, and J. K. Li, "The changes of crop yield and soil fertility with long-term fertilizer application," Plant Nutrition Fertilizer Science, vol. 1, no. 1, pp. 6–18, 1994.

37. S. C. Brubaker, A. J. Jones, D. T. Lewis, and K. Frank, "Soil properties associated with landscape position," Soil Science Society of America Journal, vol. 57, no. 1, pp. 235–239, 1993.

38. M. Reeves, R. Lal, T. Logan, and J. Sigarán, "Soil nitrogen and carbon response to maize cropping system, nitrogen source, and tillage," Soil Science Society of America Journal, vol. 61, no. 5, pp. 1387–1392, 1997.

39. C. J. Wright and D. C. Coleman, "The effects of disturbance events on labile phosphorus fractions and total organic phosphorus in the southern Appalachians," Soil Science, vol. 164, no. 6, pp. 391–402, 1999.

40. G. Lian, D. Wang, J. Lin, and D. Yan, "Characteristics of nutrient leaching from paddy field in Taihu Lake area," Chinese Journal of Applied Ecology, vol. 14, no. 11, pp. 1879–1883, 2003.

41. J. G. Xi and J. B. Zhou, "Leaching and transforming characteristics of urea N added by different ways of fertigation," Plant Nutrition Fertilizer Science, vol. 9, no. 3, pp. 271–275, 2003.

42. G. W. Thomas, "Effects of a killed sod mulch on movement and corn yield," Agronomy Journal, vol. 65, pp. 736–739, 1973.

43. R. L. Blevins, G. W. Thomas, and Cornelius, "Influence of no-tillage and nitrogen fertilization on soil properties after 5 years of continuous corn," Agronomy Journal, vol. 69, pp. 383–386, 1977.

44. D. D. Tyler and G. W. Thomas, "Lysimeter measurements of nitrate and chloride losses from soil under conventional and no tillage corn," Journal of Environmental Quality, vol. 6, no. 1, pp. 63–66, 1977.

45. C. W. Rice and M. S. Smith, "Denitrification in no-till and plowed soils," Soil Science Society of America Journal, vol. 46, no. 6, pp. 1168–1173, 1982.

46. EL-Baruni and S. R. Olsen, "Effects of manure on solubility of phosphorus in calcareous soils," Soil Science, vol. 112, pp. 219–225, 1979.

47. T. Ohno and M. S. Erich, "Inhibitory effects of crop residue-derived organic ligands on phosphate adsorption kinetics," Journal of Environmental Quality, vol. 26, no. 3, pp. 889–895, 1997.

48. G. T. Li, Z. J. Zhao, Y. F. Huang, and B. G. Li, "Effect of straw returning on soil nitrogen transformation," Plant Nutrition and Fertilizer Science, vol. 8, no. 2, pp. 162–167, 2002.

49. S. Nolan, M. Unkovich, S. Yuying, L. Lingling, and W. Bellotti, "Farming systems of the Loess Plateau, Gansu Province, China," Agriculture, Ecosystems and Environment, vol. 124, no. 1-2, pp. 13–23, 2008.

50. T. M. McCalla and T. J. Army, "Stubble Mulch Farming," Advances in Agronomy, vol. 13, no. C, pp. 125–196, 1961.

51. F .B. Ellis, J. G. Elliot, F. Pollard, R. Q. Cannell, and B. T. Barnes, "Comparison of direct drilling, reduced cultivation and ploughing on the growth of cereals: 3. Winter wheat and spring barley on calcareous clay," The Journal of Agriculture Science, vol. 93, pp. 391–401, 1979.

52. J. P. Graham, F. B. Ellis, and D. G. Christian, "Effects of straw residues on the establishment, growth and yield of autumn-sown cereals," Journal of Agriculture Engineering Research, vol. 24, pp. 39–49, 1986.

Chapter 4

Crop Diversity Effects on Near-Surface Soil Condition under Dryland Agriculture

Mark A. Liebig, David W. Archer, and Don L. Tanaka

Northern Great Plains Research Laboratory, ND 58554-0459, USA

ABSTRACT

Unprecedented changes in agricultural land use throughout the northern Great Plains of North America have highlighted the need to better understand the role of crop diversity to affect ecosystem services derived from soil. This study sought to determine the effect of four no-till cropping systems differing in rotation length and crop diversity on near-surface (0 to 10 cm) soil properties. Cropping system treatments included small grain-fallow (SG-F) and three continuously cropped rotations (3 yr, 5 yr, and Dynamic) located in south-central North Dakota,

USA. Soil pH was lower in the 3 yr rotation (5.17) compared to the Dynamic (5.51) and SG-F (5.55) rotations ($P \leq 0.05$). Among cropping system treatments, 5 yr and Dynamic rotations possessed significantly greater soil organic C (SOC) and total N (mean = 26.3 Mg C ha^{-1}, 2.5 Mg N ha^{-1}) compared to the 3 yr (22.7 Mg C ha^{-1}, 2.2 Mg N ha^{-1}) and SG-F (19.9 Mg C ha^{-1}, 2.0 Mg N ha^{-1}) rotations ($P \leq 0.05$). Comparison of SOC measured in this study to baseline values at the research site prior to the establishment of treatments revealed only the 5 yr and Dynamic rotations increased SOC over time. The results of this study suggest that a diverse portfolio of crops is necessary to minimize soil acidification and increase SOC.

INTRODUCTION

Producing a sufficient amount of food while protecting environmental quality and sustaining rural economies represents a significant agricultural challenge in the 21st century [1]. The immensity of this challenge is brought into focus when considering current trajectories in climate change and nonrenewable resource use [2]. Accordingly, increased emphasis has been placed on developing agricultural production systems that are inherently resilient to external stressors yet are highly productive, economically competitive, and environmentally benign. This nexus of productivity, profitability, and ecosystem health has underscored the critical role of soil, Earth's biogeochemical engine, to affect agricultural and environmental outcomes through impacts on ecosystem services [3].

Classification of soil ecosystem services is encompassed within supporting, regulating, provisioning, and cultural categories [4]. The retention and delivery of plant nutrients (supporting), regulation of element and hydrologic cycles (regulating), and physical support for plants (provisioning) can be inferred through the measurement of key soil physical, chemical, and biological properties and processes. Such assessments are needed to elucidate management effects on soil ecosystem services that directly affect agricultural sustainability.

Of the broad array of management decisions under direct producer control, crop rotation perhaps represents the most significant with regard to long-term economic and environmental outcomes [1, 5].

In the context of environmental outcomes related to soil ecosystem services, crop rotation effects can be manifested through alterations in soil structure, soil-water properties, and/or nutrient retention and availability.

Crop rotations including perennial legumes or grasses can increase the formation and stability of aggregates compared to two-year crop rotations or monocultures [6, 7]. Improvements in soil structure under extended crop rotations have corresponded with lower soil bulk density and higher infiltration rates [8, 9]. Crop residue inputs strongly affect soil nutrient stocks [10], thereby limiting generalizations regarding effects of extended crop rotations on soil C and N. Decreases in soil C and N have been observed with the inclusion of leguminous crops in rotation [11, 12]. Fixed N by leguminous crops, however, has been associated with greater net N mineralization in extended crop rotations compared to monocultures [9, 13]. Collectively, integrative assessments using the Soil Management Assessment Framework [14] found higher overall soil quality index values in longer and more diverse crop rotations compared to two-year crop rotations or monocultures, implying improved soil function in the former [7, 15].

This study sought to quantify effects of crop rotation on a suite of soil properties within four long-term cropping systems in south-central North Dakota, USA. The region represented by the study area has undergone an unprecedented transition in agricultural land use involving the conversion of grassland to annual crops [16]. Moreover, recent documentation of cropping patterns in the region suggests an increased prevalence of monoculture cropping [17]. These regional land use trends underscore the value of understanding crop rotation effects on soil properties that infer the status of critical soil functions.

MATERIALS AND METHODS

Site and Treatment Description

The research site was located within the Missouri Plateau approximately 6 km south of Mandan, North Dakota, USA (46°46'12"N, 100°54'57"W) on the Area IV Soil Conservation Districts (SCD) Research Farm. The site is on gently rolling uplands (0–3% slope) with a silty loess mantle overlying Wisconsin-age till. Soils at the site are dominated by a mix of Temvik and Wilton silt loams (USDA: fine-silty, mixed, superactive, frigid Typic, and Pachic Haplustolls; FAO: Calcic Siltic Chernozems). Long-term (98 yr) mean annual precipitation is 412 mm, with 79% of the total received during the growing season (April–September). Long-term mean annual temperature is 4°C, though daily averages fluctuate from < -10 °C in the winter to >20 °C in the summer.

Four field-scale cropping system treatments were established at the site between 1984 and 2001 to evaluate long-term effects of rotation length and crop diversity on crop performance, precipitation-use efficiency, and soil quality [18]. Rotations and year of establishment included (1) small grain-fallow (SG-F, 1984), where the small grain included spring wheat (Triticum aestivum L.), barley (Hordeum vulgare L.), or oat (Avena sativa L.); (2) spring wheat, winter wheat (Triticum aestivum L.), and sunflower (Helianthus annuus L.) (3 yr; 1984); (3) spring wheat, winter wheat, dry pea (Pisum sativum L.), corn (Zea mays L.), and soybean (Glycine max L.) (5-yr; 2001); and (4) a Dynamic rotation that included six of the following crops: corn, sunflower, spring wheat, winter wheat, soybean, and buckwheat (Fagopyrum esculentum Moench) (Dynamic; 2001). Crops in the Dynamic rotation were sequenced each year based on market opportunities, soil water and nutrient conditions at planting, and/or restrictions on herbicide use within the planted field [19]. Each phase of the SG-F, 3 yr, and 5 yr rotations was present every year. Crops in the Dynamic rotation were also present every year, though due to the availability of seven fields for the rotation some crops were present in duplicate. Field size for each crop phase varied, ranging from 2 to 14 ha. Crop rotations were not replicated. Prior to establishment in 2001, cropland areas allocated to the 5 yr and Dynamic rotations were cropped to the 3 yr rotation.

Corn, soybean, and sunflower were planted with a John Deere MaxEmerge II row crop planter in 76 cm rows (Deere & Company, Moline, IL), while all other crops were planted in 19 cm rows with a John Deere Model 750 no-till drill or a Bourgault air seeder (Bourgault Industries, St. Brieux, Saskatchewan, Canada). Nitrogen fertilizer (ammonium nitrate or urea) was applied to all crops prior to or concurrent with planting operations at recommended rates [20] while taking into consideration levels of residual soil N following years with below normal precipitation. Phosphorus fertilizer (triple superphosphate) was applied annually to all crops at 11 kg P ha^{-1} with the seed at planting. Burn-down and postemergent herbicides were applied to crops in all rotations to control weeds as needed, as were fungicides to control leaf spot disease. Scheduling of seeding, fertilizer and pesticide application, and harvest followed best management practices used by area producers.

Sampling Protocol and Laboratory Analyses

In May 2012, three $10 m^2$ pseudoreplicates were established in each crop rotation phase prior to planting, resulting in 51 pseudoreplicates across all treatments (6, 9, 15, and 21 pseudoreplicates for SG-F, 3 yr, 5 yr, and Dynamic rotations, respectively). Selection of pseudoreplicates in each treatment was done carefully to ensure all sampling sites possessed the same soil type (Temvik silt loam) and landscape attributes (0-1% slope). Eight soil cores were collected in each pseudoreplicate from the 0 to 10 cm depth using a 3.13 cm (internal diameter) step-down probe and composited. To ensure composite samples were representative of each plot, three cores each were collected from the nonwheel- and wheel-tracked interrows, and two cores from the row. Each sample was stored in a double-lined plastic bag, placed in cold storage at 5°C, and analyzed for chemical and biological attributes within 3 weeks of collection.

Infiltration rate measurements (one pseudoreplicate^{-1}) were made at the time of sampling by inserting a piece of heavy-gauge aluminum irrigation pipe (15 cm internal diameter by 15 cm length) into the soil of a nonwheel-trafficked interrow to a 7.5 cm depth and applying two separate applications of water within the enclosed space of the ring [21]. The volume of water for each application was equivalent to a 2.54 cm depth in the ring. The time necessary for each application of water to infiltrate into the soil was recorded using a stopwatch. To eliminate effects associated with differences in antecedent water content among treatments, only data from the second water application were analyzed.

Soil processing was initiated by weighing the total tared soil mass at field moisture content. Gravimetric water content was determined for each sample by removing a 12–15 g subsample and measuring the difference in mass before and after drying at 105°C [22]. Samples were then split for chemical and biological analyses into two approximately equal portions. Samples for chemical analyses were dried at 32°C for 3 to 4 d and then ground by hand to pass a 2.0 mm sieve. Identifiable plant material (>2.0 mm diameter, >10 mm length) was removed during sieving and discarded. Chemical analyses included assessments of electrical conductivity (EC), soil pH, particulate organic matter (POM), and total C and N. Soil pH and EC were estimated from a 1:1 soil-water mixture [23, 24]. Particulate organic matter (POM) was quantified by analyzing the C content of material retained on a 0.053 mm sieve [25]. Particulate organic matter C, along with total soil C and N, was determined by dry combustion. As pH was <7.2 for the depths sampled, total soil C was considered equivalent to soil organic C (SOC).

Biological analyses included assessment of soil microbial biomass, which was estimated using the microwave irradiation method [26]. Prior to analysis, each split sample was sieved through a 2.0 mm sieve at field moisture content. Fifty grams of sieved soil was incubated 10 d at 55% water-filled pore space in the presence of 10 mL of 2.0 M NaOH. Carbon dioxide content was determined by single end-point titration with 0.1 M HCl [27], and the flush of CO_2-C following irradiation was calculated without subtracting a 10 d control [28]. Gravimetric data were converted to a volumetric basis using field measured soil bulk density [29]. All data were expressed on an oven-dry basis.

For purposes of comparison, near-surface soil samples (0 to 10 cm) from a grazed pasture were collected and analyzed following sampling guidelines and laboratory analyses outlined above. The pasture, located approximately 2.5 km east of the crop rotation treatments, possessing the same soil type and landscape attributes, has never been tilled and has been grazed by cattle at a low stocking rate (2.6 ha steer^{-1}) as part of a long-term experiment established in 1916 [30].

Data Analyses

Crop rotation effects on soil properties were evaluated by ANOVA using PROC mixed in SAS [31]. The PDIFF option of the LSMEANS statement was used to document differences between treatment means using a significance criterion of $P \leq 0.05$. Means were calculated across phases for each crop rotation. Means of soil properties within the grazed pasture were not included in data analyses but presented for general comparison only.

RESULTS AND DISCUSSION

Crop rotation diversity had a pronounced effect on soil physical condition, soil solution chemistry, and soil organic matter attributes. Soil bulk density was significantly lower in the 5 yr rotation compared to the SG-F and Dynamic rotations (Table 1), though observed values for all treatments were not indicative of physical conditions restrictive of root growth [32]. Soil bulk density among cropped treatments ranged from 0.30 to 0.37 Mg m^{-3} greater than grazed pasture, the result of abundant near-surface root biomass in the latter [33]. Despite observed differences in soil bulk density, crop rotation effects on infiltration rate were not significant (P = 0.16). Treatment assessments of infiltration rate are often challenging at large spatial scales, with coefficients of variation >100% for many water transport properties [34, 35]. Despite this fact, there was a notable numerical trend among treatments (SG-F < 3 yr < 5 yr < Dynamic), with infiltration rate increasing 6.1 to 7.1 cm hr^{-1} with increasing rotational diversity (Table 1).

Table 1: Crop diversity effects on soil bulk density at 0–10 cm and infiltration rate for long-term crop and pasture treatments near Mandan, ND

Soil property	Small grain-fallow	3 yr fixed rotation	5 yr fixed rotation	Dynamic rotation	Grazed pasture†
Bulk density (Mg m−3)	1.20 (0.02) a‡	1.18 (0.04) ab	1.13 (0.01) b	1.18 (0.01) a	0.83 (0.02)
Infiltration rate (cm hr−1)	7.9 (4.8)	15.0 (6.9)	22.1 (4.6)	28.2 (5.7)	—

†Soil property values for grazed pasture with the same soil type are shown for comparison but were not included in statistical analyses. Infiltration rate not measured in grazed pasture due to excessive soil wetness at time of sampling.

‡Values in parentheses represent mean standard error. Mean values in a row followed by a different letter are significantly different at $P \leq 0.05$.

Measurements of soil solution chemistry suggested near-surface soil conditions among crop rotation treatments were nonsaline and moderately to strongly acidic [36] (Table 2). While crop rotation did not affect EC (P = 0.15), soil pH was significantly higher in the

SG-F and Dynamic rotations compared to the 3 yr rotation. Use of urea and ammonium nitrate fertilizers, coupled with differences in N fertilization frequency, likely contributed to observed treatment effects on soil pH. Nitrification of ammonium-based fertilizers causes soil acidification, particularly if nitrate is not taken up by plant roots [37]. Such acidification is expected to be greater where N is applied each year compared to crop-fallow, where N is applied biannually. Accordingly, soil pH has been found to be higher in dryland cropping systems including fallow [38]. Decreased acidification in the Dynamic versus the 3 yr rotation was likely the result of differences in applied N over time. Compared to grazed pasture, crop rotation treatments were 0.67 to 1.05 pH units more acidic (Table 2). Moreover, soil samples collected in 1983 prior to the establishment of rotation treatments on one of the fields included in this study possessed a pH of 6.4 for the 0 to 7.6 cm depth [39], suggesting substantial surface acidification during the intervening 29 years.

Table 2: Crop diversity effects on electrical conductivity and soil pH at 0–10 cm for long-term crop and pasture treatments near Mandan, ND

Soil property	Small grain-fallow	3 yr fixed rotation	5 yr fixed rotation	Dynamic rotation	Grazed pasture†
Electrical conductivity (dS m−1)	0.59 (0.14)	0.58 (0.03)	0.66 (0.06)	0.74 (0.04)	0.39 (0.01)
Soil pH (−log[H+])	5.55 (0.16) a‡	5.17 (0.06) b	5.37 (0.04) ab	5.51 (0.09) a	6.22 (0.03)

†Soil property values for grazed native vegetation with the same soil type are shown for comparison but were not included in statistical analyses.

‡Values in parentheses represent mean standard error. Mean values in a row followed by a different letter are significantly different at $P \leq 0.05$.

Quantifying the status and trajectory of soil organic matter attributes is critically important for understanding management impacts on the productivity and stability of agroecosystems [40]. Soil organic C and total N followed a similar trend among cropped treatments, with 5 yr > Dynamic > 3 yr > SG-F (Table 3). Statistically, SOC and total N were greater in 5 yr and Dynamic rotations compared to 3 yr and

SG-F rotations, while the 3 yr rotation possessed significantly greater total N than SG-F ($P \leq 0.01$). Soil C:N ratio for cropped treatments ranged between 10.20 and 10.73, with values greatest for the 5 yr and Dynamic rotations (10.67 to 10.73) and least for the 3 yr and SG-F rotations (10.20 to 10.23). Statistical differences in C:N ratio among cropped treatments were limited to 5 yr > 3 yr and SG-F and 3 yr < 5 yr and Dynamic ($P = 0.02$). Particulate organic matter C, a moderately labile fraction composed mostly of plant residue, exhibited the widest range in observed values among C and N parameters (1172 to 3078 kg C ha^{-1}). Among cropped treatments, POM-C was the greatest in the 5 yr rotation, intermediate in the Dynamic rotation, and the least in SG-F ($P \leq 0.01$). Particulate organic matter in the 3 yr rotation was not different from the 5 yr and Dynamic rotations. Microbial biomass C was not different among cropped treatments ($P = 0.17$).

Table 3: Crop diversity effects on soil carbon, nitrogen, C:N ratio, particulate organic matter (POM) C, and microbial biomass C at 0–10 cm for long-term crop and pasture treatments near Mandan, ND

Soil property	Small grain-fallow	3 yr fixed rotation	5 yr fixed rotation	Dynamic rotation	Grazed pasture[†]
Soil organic C (Mg C ha−1)	19.9 (1.0) b‡	22.7 (0.9) b	26.9 (0.4) a	25.7 (0.7) a	31.0 (0.7)
Total N (Mg N ha−1)	2.0 (0.1) c	2.2 (0.1) b	2.5 (0.1) a	2.4 (0.1) a	2.6 (0.1)
C:N ratio	10.23 (0.26) bc	10.20 (0.11) c	10.73 (0.10) a	10.67 (0.12) ab	12.02 (0.11)
POM-C (kg C ha−1)	1172 (144) c	2834 (151) ab	3078 (162) a	2486 (144) b	4690 (262)
Microbial biomass C (kg C ha−1)	549 (48)	748 (89)	657 (42)	671 (32)	992 (116)

[†]Soil property values for grazed native vegetation with the same soil type are shown for comparison but were not included in statistical analyses.

[‡]Values in parentheses represent mean standard error. Mean values in a row followed by a different letter are significantly different at $P \leq 0.05$.

All soil C and N parameters were numerically greater in grazed pasture than cropped treatments. Soil organic C, total N, C:N ratio, POM-C, and MBC were 15, 4, 12, 52, and 33% greater, respectively, in grazed pasture compared to the cropped treatment with the highest value for each parameter (5 yr rotation for four of five parameters). In contrast, the same five parameters were 56, 30, 17, 300, and 81% greater in grazed pasture compared to SW-F.

Crop rotations contributing greater above- and below-ground biomass generally increase soil C and N under conditions of equivalent tillage and nutrient management [5]. Accordingly, inclusion of fallow in semiarid cropping systems is associated with decreased C and N in near-surface soil depths compared to continuous cropping due to lower biomass contributions in the former [41, 42]. Under continuous cropping, effects of crop rotation length and/or crop diversity on soil C and N have been mixed depending on residue quantity and quality, as well as water availability for growth of subsequent crops [43]. In this study, crops included in the 5 yr and Dynamic rotations were sequenced to favor snow capture and efficient precipitation use, which would serve to enhance production over a rotation cycle [44]. While the inclusion of wheat was a consistent feature in the continuously cropped rotations, it is possible that the addition of corn, a high residue-producing crop with a moderately high C:N stover ratio, may have contributed to increased SOC and TN in the 5 yr and Dynamic rotations. Sequencing corn after dry pea (as done in the 5 yr rotation) has been found to increase corn residue production by 33 to 55% compared to corn after corn [45], thereby providing increased aboveground biomass in the former, which serves as an important precursor to SOC accrual [41]. Sherrod et al. [46, 47] observed greater SOC and POM-C under continuous corn compared to wheat-corn-fallow and wheat-fallow in a long-term study in eastern Colorado. Conversely, sunflower, a crop well-known for high water use and limited residue production relative to other crops common to the northern Plains [44, 48], can severely restrict crop production in subsequent years when precipitation is below the long-term mean [49]. Given its inclusion in the 3 yr rotation coupled with the consistency of periodic drought in the region, constraints to spring wheat production would be expected during drought years following sunflower [50].

Comparison of soil C and N pools between grazed pasture and cropped treatments indirectly reflected effects of cropping system diversity relative to a "native" baseline. Previous comparisons of cropland and virgin grassland in southwest North Dakota documented differences in SOC ranging from 33 to 36% in the 0 to 15.2 cm depth [51, 52]. Accordingly, results from this study suggesting modern cropping systems utilizing diverse rotations under no-till management have narrowed the gap in near-surface SOC between cropland and un-tilled native grassland. This inference is predicated on the assumption that SOC in native grassland has not decreased, which recent assessments suggest is not the case [53]. This inference is further supported by a previous evaluation of soil conditions at the research site. Black and Tanaka [39] reported a SOC content of $21.4\,g\,C\,kg^{-1}$ at 0 to 7.6 cm and $20.5\,g\,C\,kg^{-1}$ at 7.6 to 15.2 cm in 1983 prior to the establishment of rotation treatments. When these values were weighted to a 0 to 10 cm depth and compared to SOC measured in this study, the 5 yr and Dynamic rotations were found to have increased by 2.6 and $0.6\,g\,C\,kg^{-1}$, respectively. Conversely, SOC in the SW-F and 3 yr rotations decreased by 4.6 and $2.0\,g\,C\,kg^{-1}$ compared to baseline measurements in 1983. Such findings suggest accrual of near-surface SOC in northern Plains no-till cropping systems requires not just continuous cropping, but a diverse mixture of crops sequenced in a manner to enhance biomass production over the long-term.

CONCLUSIONS

Increased crop diversity has been found to foster greater and more stable crop yields, improved nutrient- and water-use efficiencies, and increased profit compared to less diverse cropping systems [1, 45, 54]. Less attention, however, has been directed to understanding crop diversity effects on soil properties, particularly in the northern Great Plains of North America.

Under conditions of this study, decreased soil acidification in the Dynamic and SW-F rotations compared to the 3 yr rotation implied greater resistance to pH change in the former, though mechanisms for resistance likely differ between rotations. In SW-F, N fertilizer was applied biannually, thereby decreasing the rate of acidification from N loss compared to continuously cropped rotations where fertilizer N was applied annually. In contrast, crops in the Dynamic rotation were sequenced to optimize nutrient and precipitation use which, over the long-term, would serve to reduce N loss. A detailed characterization of N use efficiency among cropped treatments is needed to ascertain the suitability of this inferred mechanism in the Dynamic rotation.

Among continuously cropped treatments, soil organic matter attributes were generally the greatest in the 5 yr rotation, intermediate in the Dynamic rotation, and the least in the 3 yr rotation. Moreover, comparison of SOC values measured in this study to baseline values measured prior to the establishment of rotation treatments indicated only the 5 yr and Dynamic rotations increased SOC in the near-surface depth over time. These findings suggest rotating a diverse portfolio of annual crops under no-till management is necessary to maintain or accrue SOC in this region.

ACKNOWLEDGMENTS

The authors acknowledge the contribution of the Area IV SCD in North Dakota for providing land to conduct research reported in this paper. Marvin Hatzenbuhler managed cropping system treatments, Branden Bott and Anna Hruby assisted with field sampling and soil processing, and Johannah Miller and Becky Wald conducted laboratory assessments. They also acknowledge the many field and laboratory support personnel who have invested countless hours to maintain plots and collect data on the Area IV SCD Research Farm since its inception. The U.S. Department of Agriculture, Agricultural Research Service, is an equal opportunity/affirmative action employer and all agency services are available without discrimination. Mention of commercial products and organizations in this paper is solely to provide specific information. It does not constitute endorsement by USDA-ARS over other products and organizations not mentioned.

REFERENCES

1.   A. S. Davis, J. D. Hill, C. A. Chase, A. M. Johanns, and M. Liebman, "Increasing cropping system diversity balances productivity, profita bility and environmental health," PLoS ONE, vol. 7, no. 10, Article ID e47149, 2012.

2. J. A. Foley, N. Ramankutty, K. A. Brauman et al., "Solutions for a cultivated planet," Nature, vol. 478, no. 7369, pp. 337–342, 2011.

3. D. A. Robinson, N. Hockley, E. Dominati et al., "Natural capital, ecosystem services, and soil change: why soil science must embrace an ecosystems approach," Vadose Zone Journal, vol. 11, no. 1, 2012.

4. G. C. Daly, P. A. Matson, and P. M. Vitousek, "Ecosystem services supplied by soils," in Natures Services: Societal Dependence on Natural Ecosystems, G. C. Daly, Ed., pp. 113–132, Island Press, Washington, DC, USA, 1997.

5. D. L. Karlen, G. E. Varvel, D. G. Bullock, and R. M. Cruse, "Crop Rotations for the 21st Century,"Advances in Agronomy, vol. 53, pp. 1–45, 1994.

6. B. A. Raimbault and T. J. Vyn, "Crop rotation and tillage effects on corn growth and soil structural stability," Agronomy Journal, vol. 83, pp. 979–985, 1991.

7. D. L. Karlen, E. G. Hurley, S. S. Andrews et al., "Crop rotation effects on soil quality at three northern corn/soybean belt locations," Agronomy Journal, vol. 98, no. 3, pp. 484–495, 2006.

8. S. D. Logsdon, J. K. Radke, and D. L. Karlen, "Comparison of alternative farming systems. I. Infiltration techniques," American Journal of Alternative Agriculture, vol. 8, no. 1, pp. 15–20, 1993.

9. A. E. Russell, D. A. Laird, and A. P. Mallarino, "Nitrogen fertilization and cropping system impacts on soil quality in Midwestern Mollisols," Soil Science Society of America Journal, vol. 70, no. 1, pp. 249–255, 2006.

10. W. W. Wilhelm, J. M. F. Johnson, D. L. Karlen, and D. T. Lightle, "Corn stover to sustain soil organic carbon further constrains biomass supply," Agronomy Journal, vol. 99, no. 6, pp. 1665–1667, 2007.

11. D. R. Huggins, R. R. Allmaras, C. E. Clapp, J. A. Lamb, and G. W. Randall, "Corn-soybean sequence and tillage effects on soil carbon dynamics and storage," Soil Science Society of America Journal, vol. 71, no. 1, pp. 145–154, 2007.

12. G. E. Varvel and W. W. Wilhelm, "Soil carbon levels in irrigated western corn belt rotations," Agronomy Journal, vol. 100, no. 4, pp. 1180–1184, 2008.

13. M. A. Liebig, G. E. Varvel, J. W. Doran, and B. J. Wienhold, "Crop sequence and nitrogen fertilization effects on soil properties in the Western Corn Belt," Soil Science Society of America Journal, vol. 66, no. 2, pp. 596–601, 2002.

14. S. S. Andrews, D. L. Karlen, and C. A. Cambardella, "The soil management assessment framework: a quantitative soil quality evaluation method," Soil Science Society of America Journal, vol. 68, no. 6, pp. 1945–1962, 2004.

15. B. J. Wienhold, J. L. Pikul, M. A. Liebig et al., "Cropping system effects on soil quality in the great plains: synthesis from a regional project," Renewable Agriculture and Food Systems, vol. 21, no. 1, pp. 49–59, 2006.

16. K. C. Wright and M. C. Wimberly, "Recent land use change in the Western Corn Belt threatens grasslands and wetlands," Proceedings of the National Academy of Sciences of the United States of America, vol. 110, no. 10, pp. 4134–4139, 2013.

17. J. D. Plourde, B. C. Pijanowski, and B. K. Pekin, "Evidence for increased monoculture cropping in the Central United States," Agriculture, Ecosystems and Environment, vol. 165, pp. 50–59, 2013.

18. D. L. Tanaka and M. A. Liebig, "Crop rotation pri nciples for the northern Plains," in Proceedings of the 34th Annual Zero Tillage Workshop and Trade Show, pp. 31–32, Manitoba—North Dakota Zero Tillage Farmers Association, Minot, ND, USA, January 2012.

19. D. L. Tanaka, J. M. Krupinsky, M. A. Liebig et al., "Dynamic cropping systems: an adaptable approach to crop production in the Great Plains," Agronomy Journal, vol. 94, no. 5, pp. 957–961, 2002.

20. NDSU, Fertilizer: Recommendations, North Dakota State University Extension Service, 2014,http://www.ag.ndsu.edu/publications/crops/soil-fertilizer.

21. M. Sarrantonio, J. W. Doran, M. A. Liebig, and J. J. Halvorson, "On-farm assessment of soil quality and health," in Methods for Assessing Soil Quality, J. W. Doran and A. J. Jones, Eds., pp. 83–105, SSSA, Madison, Wis, USA, 1996.

22. W. H. Gardner, "Water content," in Methods of Soil Analysis. Part 1—Physical and Mineralogical Methods, A. Klute, Ed., Soil

Science Society of America Book Series No. 5, pp. 493–544, Soil Science Society of America and American Society of Agromy, Madison, Wis, USA, 2nd edition, 1986.

23. M. E. Watson and J. R. Brown, "pH and lime requirement," in Recommended Chemical Soil Test Procedures for the North Central Region, J. R. Brown, Ed., vol. 221 of Missouri Agriculture Experiment Station Bulletin, pp. 13–16, North Central Region Publication, 1998.

24. D. A. Whitney, "Soil salinity," in Recommended Chemical Soil Test Procedures for the North Central Region, J. R. Brown, Ed., vol. 221 of North Central Regional Publications, pp. 59–60, Missouri Agriculture Experiment Station Bulletin, 1998.

25. E. G. Gregorich and B. H. Ellert, "Light fraction and macroorganic matter in mineral soils," in Soil Sampling Methods and Analysis, M. R. Carter, Ed., pp. 397–407, Canadian Society of Soil Science, Lewis, Boca Raton, Fla, USA, 1993.

26. K. R. Islam and R. R. Weil, "Microwave irradiation of soil for routine measurement of microbial biomass carbon," Biology and Fertility of Soils, vol. 27, no. 4, pp. 408–416, 1998.

27. E. A. Paul, D. Harris, M. J. Klug, and R. W. Ruess, "The determination of microbial biomass," inStandard Soil Methods for Long-Term Ecological Research, G. P. Robertson, D. C. Coleman, C. S. Bledsoe, and P. Sollins, Eds., pp. 291–317, Oxford University Press, New York, NY, USA, 1999.

28. A. J. Franzluebbers, R. L. Haney, F. M. Hons, and D. A. Zuberer, "Assessing biological soil quality with chloroform fumigation-incubation: why subtract a control?" Canadian Journal of Soil Science, vol. 79, no. 4, pp. 521–528, 1999.

29. G. R. Blake and K. H. Hartge, "Bulk density," in Methods of Soil Analysis. Part 1—Physical and Mineralogical Methods, A. Klute, Ed., Soil Science Society of America Book Series No. 5, pp. 363–382, Soil Science Society of America and American Society of Agromy, Madison, Wis, USA, 2nd edition, 1986.

30. J. T. Sarvis, "Effects of different systems and intensities of grazing upon the native vegetation at the Northern Great Plains field station," USDA, Department Bulletins 1170, U.S. Government Printing Office, Washington, Wash, USA, 1923.

31. R. C. Littell, G. A. Milliken, W. W. Stroup, and R. D. Wolfinger, SAS System for Mixed Models, SAS Institute, Cary, NC, USA, 1996.

32. C. A. Jones, "Effect of soil texture on critical bulk densities for root growth," Soil Science Society of America Journal, vol. 47, no. 6, pp. 1208–1211, 1983.

33. A. B. Frank, M. A. Liebig, and D. L. Tanaka, "Management effects on soil CO_2 efflux in northern semiarid grassland and cropland," Soil and Tillage Research, vol. 89, no. 1, pp. 78–85, 2006.

34. J. L. Pikul, R. C. Schwartz, J. G. Benjamin, R. L. Baumhardt, and S. Merrill, "Cropping system influences on soil physical properties in the Great Plains," Renewable Agriculture and Food Systems, vol. 21, no. 1, pp. 15–25, 2006.

35. S. D. Logsdon, R. R. Allmaras, L. Wu, J. B. Swan, and G. W. Randall, "Macroporosity and its relation to saturated hydraulic conductivity under different tillage practices," Soil Science Society of America Journal, vol. 54, no. 4, pp. 1096–1101, 1990.

36. USDA, Soil Survey Manual, Handbook No. 18, USDA Soil Survey Division Staff. U.S. Dep. of Agric, U.S. Govt. Printing Office, Washington, DC, USA, 1993.

37. O. T. Bouman, D. Curtin, C. A. Campbell, V. O. Biederbeck, and H. Ukrainetz, "Soil acidification from long-term use of anhydrous ammonia and urea," Soil Science Society of America Journal, vol. 59, no. 5, pp. 1488–1494, 1995.

38. M. A. Liebig, D. L. Tanaka, and B. J. Wienhold, "Tillage and cropping effects on soil quality indicators in the Northern Great Plains," Soil and Tillage Research, vol. 78, no. 2, pp. 131–141, 2004.

39. A. L. Black and D. L. Tanaka, "A conservation tillage-cropping systems study in the northern Great Plains of the United States," in Soil Organic Matter in Temperate Agroecosystems, E. A. Paul, K. Paustian, E. T. Elliott, and C. V. Cole, Eds., pp. 335–342, CRC Press, Boca Raton, Fla, USA, 1997.

40. H. H. Janzen, "Soil carbon: a measure of ecosystem response in a changing world?" Canadian Journal of Soil Science, vol. 85, no. 4, pp. 467–480, 2005.

41. H. P. Collins, M. M. Mikha, T. T. Brown, J. L. Smith, D. Huggins, and U. M. Sainju, "Agricultural management and soil carbon dynamics: Western U.S. croplands," in Managing Agricultural Greenhouse Gases: Coordinated Agricultural Research through GRACEnet to Address our Changing Climate, M. A. Liebig, A. J. Franzluebbers, and R. F. Follett, Eds., pp. 59–77, Academic Press, San Diego, Calif, USA, 2012.

42. A. J. VandenBygaart, E. G. Gregorich, and D. A. Angers, "Influence of agricultural management on soil organic carbon: a compendium and assessment of Canadian studies," Canadian Journal of Soil Science, vol. 83, no. 4, pp. 363–380, 2003.

43. M. M. Mikha, J. G. Benjamin, M. F. Vigil, and D. C. Nielson, "Cropping intensity impacts on soil aggregation and carbon sequestration in the central great plains," Soil Science Society of America Journal, vol. 74, no. 5, pp. 1712–1719, 2010.

44. S. D. Merrill, D. L. Tanaka, J. M. Krupinsky, M. A. Liebig, and J. D. Hanson, "Soil water depletion and recharge under ten crop species and applications to the principles of dynamic cropping systems,"Agronomy Journal, vol. 99, no. 4, pp. 931–938, 2007.

45. D. L. Tanaka, J. M. Krupinsky, S. D. Merrill, M. A. Liebig, and J. D. Hanson, "Dynamic cropping systems for sustainable crop production in the Northern Great Plains," Agronomy Journal, vol. 99, no. 4, pp. 904–911, 2007.

46. L. A. Sherrod, L. R. Ahuja, G. A. Peterson, and D. G. Westfall, "Soil organic carbon pools after 12 years in no-till dryland agroecosystems," Soil Science Society of America Journal, vol. 69, no. 5, pp. 1600–1608, 2005.

47. L. A. Sherrod, G. A. Peterson, D. G. Westfall, and L. R. Ahuja, "Cropping intensity enhances soil organic carbon and nitrogen in a no-till agroecosystem," Soil Science Society of America Journal, vol. 67, no. 5, pp. 1533–1543, 2003

48. J. M. Krupinsky, S. D. Merrill, D. L. Tanaka, M. A. Liebig, M. T. Lares, and J. D. Hanson, "Crop residue coverage of soil influenced by crop sequence in a no-till system," Agronomy Journal, vol. 99, no. 4, pp. 921–930, 2007.

49. D. L. Tanaka, M. A. Liebig, J. M. Krupinsky, and S. D. Merrill, "Crop sequence influences on sustainable spring wheat production in the northern Great Plains," Sustainability, vol. 2, no. 12, pp. 3695–3709, 2010

50. R. A. Bowman, D. C. Nielsen, M. F. Vigil, and R. M. Aiken, "Effects of sunflower on soil quality indicators and subsequent wheat yield," Soil Science, vol. 165, no. 6, pp. 516–522, 2000

51. A. Bauer and A. L. Black, "Soil carbon, nitrogen, and bulk density comparisons in two cropland tillage systems after 25 years and in virgin grassland," Soil Science Society of America Journal, vol. 45, pp. 1166–1170, 1981.

52. H. J. Hass, C. E. Evans, and E. F. Miles, "Nitrogen and carbon changes in Great Plains soils as influenced by cropping and soil treatments," Tech. Bull. No. 1164, USDA, U.S. Government Printing Office, Washington, DC, USA, 1957.

53. M. A. Liebig, J. R. Gross, S. L. Kronberg, R. L. Phillips, and J. D. Hanson, "Grazing management contributions to net global warming potential: a long-term evaluation in the Northern great plains,"Journal of Environmental Quality, vol. 39, no. 3, pp. 799–809, 2010.

54. R. L. Anderson, R. A. Bowman, D. C. Nielsen, M. F. Vigil, R. M. Aiken, and J. G. Benjamin, "Alternative crop rotations for the central Great Plains," Journal of Production Agriculture, vol. 12, no. 1, pp. 95–99, 1999.

Chapter 5

The Role of Biochar in Ameliorating Disturbed Soils and Sequestering Soil Carbon in Tropical Agricultural Production Systems

Wolde Mekuria[1] and Andrew Noble[2]

[1]International Water Management Institute (IWMI), Addis Ababa, Ethiopia

[2]International Water Management Institute (IWMI), 127 Sunil Mawatha, Pelawatte, Battaramulla, Colombo, Sri Lanka

ABSTRACT

Agricultural soils in the tropics have undergone significant declines in their native carbon stock through the long-term use of extractive farming practices. However, these soils have significant capacity to sequester CO_2 through the implementation of improved land management practices. This paper reviews the published and grey literature related to the influence of improved land management practices on soil carbon stock in the tropics. The review suggests that the implementation of improved land management practices such as crop rotation, no-till, cover crops, mulches, compost, or manure can be effective in enhancing soil organic carbon pool and agricultural productivity in the tropics. The benefits of such amendments were, however, often short-lived, and the added organic matters were usually mineralized to CO_2 within a few cropping seasons leading to large-scale leakage. We found that management of black carbon (C), increasingly referred to as biochar, may overcome some of those limitations and provide an additional soil management option. Under present circumstances, recommended crop and land management practices are inappropriate for the vast majority of resource constrained smallholder farmers and farming systems. We argue that expanding the use of biochar in agricultural lands would be important for sequestering atmospheric CO_2 and mitigating climate change, while implementing the recommended crop and land management practices in selected areas where the smallholder farmers are not resource constrained.

INTRODUCTION

Evidence from the Intergovernmental Panel on Climate Change [1] is now overwhelmingly convincing that climate change is real, that it will intensify, and that the poorest and most vulnerable will be disproportionately affected by these changes. Climate change and variability, drought, and other climate-related extremes have a direct influence on the quantity and quality of agricultural production and, in many cases, adversely affect it. In particular, the influences of climate change on agricultural production are severe in developing countries because the technology generation, innovation, and adoption are too slow to counteract the adverse effects of varying and changing environmental conditions [2].

Agricultural intensification invariably has several negative impacts on the environment [3]. One of the major consequences of agricultural intensification is a transfer of carbon (C) to the atmosphere in the form of carbon dioxide (CO_2), thereby reducing ecosystem C pools.

Agriculture contributes 10–12% of the total global anthropogenic greenhouse gas emissions [4, 5]. Tropical agricultural soils in particular have undergone significant depletion of their native carbon stocks but have considerable potential to act as CO_2 sinks through improved management practices [6, 7]. To this end, locally appropriate adaptation and mitigation strategies to increasing climate variability and climate change are urgently needed especially in vulnerable regions where food and fiber production are most sensitive to climatic fluctuations [2].

The implementation of improved land management practices to build up carbon stocks in terrestrial ecosystems is a proven technology in reducing the concentration of CO_2 in the atmosphere and lowering atmospheric CO_2 [8]. As a result, soil organic carbon sequestration in agricultural lands has recently drawn attention in mitigating increases in atmospheric CO_2 concentrations [5, 9].

Management practices to build up soil carbon must increase the input of organic matter to soil and/or decrease soil organic matter decomposition rate. At this point, it is worth to mention that the most appropriate management practices to increase soil carbon vary regionally, depending on both environmental and socioeconomic factors. In the tropics, increasing carbon inputs through improving the fertility and productivity of crop land and pasture is essential because climate change has negative influence on the livelihood of the vast majority smallholder farmers. In extensive systems with vegetative fallow period, planted fallows and cover crops can increase carbon levels over the cropping cycle [10]. Uses of no-till, green manures, and agroforestry are other beneficial practices to sequester soil C. Overall, improving the productivity and sustainability of existing agricultural lands is crucial to reduce the rate of new land clearing and the amounts of CO_2 emitted to the atmosphere.

Carbon sequestration in agricultural soils is frequently promoted as a practical solution to slow down the rate of increase of CO_2 in the atmosphere. To date, the bulk of research into agricultural production systems being net carbon sinks has been based on temperate based production systems that are quite different to those in the tropics. There is a need to improve our understanding on how land management practices affect exchange processes that lead to net removal of atmospheric CO_2 in the tropics. Therefore, we reviewed and analyzed the impacts of different land management practices such as agronomic practices, tillage, organic input management (i.e., addition of compost, manure, biochar, and other clay materials), and agroforestry as a means to increase carbon sequestration in agricultural soils of the tropics. Moreover, we reviewed the opportunities and constraints to adapt mitigation and adaptation options in the tropics with the goal of developing a possible framework for smallholder farmers to benefit from carbon markets.

OVERVIEW OF TROPICAL PRODUCTION SYSTEMS

In the tropics, farming systems have undergone major changes from hunting and gathering through fallowing to stationary cultivation systems during the course of history [11]. The changes in farming systems were mainly driven by the increase in human population and agricultural mechanization (e.g., [12]), the quality and availability of land resources (e.g., [13]), and access to markets (e.g., [12]). The decision-making environment in tropical agricultural systems is complex. The first factor that controls farmer's decision-making is the physical environment (e.g., [14]). The second can be called household characteristics (e.g., [15]), where uniform producers react in a uniform and rational manner during decision-making processes. The third factor that determine farmers' choice of production systems is the politico/economical frame conditions (e.g., [16,17]).

Furthermore, farming systems may be differentiated into subtypes that continue to evolve along different pathways. For example, in systems under population and market pressure, some farms could successfully intensify and even specialize to produce for the market, whereas others could regress to low-input/low-output systems [17]. Moreover, in any one location within a farming system, different farms are likely to be at different stages of evolution because of differentiated resource bases, household goals, capacity to bear risk, and degree of market access, among others [15]. It is the sum of all farmers' decisions that will determine the quality of the future land resources in the tropics as there is a feedback mechanism between farmer's decisions and the quality of the natural resource base. The chain of linkages among population pressure, agricultural intensification, economic growth, societal well-being, and technical changes in agriculture and their subsequent environmental consequences are thus very complex in nature and causation and as a result are often difficult to analyze and understand.

TROPICAL AGRICULTURAL PRODUCTION SYSTEMS AND SOIL CARBON

The key problem of tropical agriculture is the steady decline in soil fertility, which is due primarily to soil erosion and the loss of soil organic matter. Some soils in tropical agricultural systems are estimated to have lost as much as 20 to 80 t Cha^{-1}, most of which has been released into the atmosphere [3, 18]. The low soil organic carbon content is due to the low shoot and root growth of crops and natural vegetation, the rapid turnover rates of organic material as a result of high soil temperatures and fauna activity particularly termites, and the low soil clay content [19]. In addition, soil erosion and the long-term cultivation using conventional tillage practices reduce soil carbon levels, and over time the soils have become degraded, often resulting in land abandonment [6, 20, 21]. For instance, a study in west African agro-ecosystem revealed that there is a tremendous SOC loss in agricultural lands due to soil erosion (ranging from 65 to 1801kgha^{-1}yr^{-1}) compared to the loss of SOC (ranging from 6 to 13kgha^{-1}yr^{-1}) in undisturbed ecosystems [19]. The loss of SOC could range from 9 to 65% depending on severity of soil erosion and soil types.

Furthermore, the burning of biomass or vegetation as a conventional land preparation method and the use of crop residues and cow dung as a source of energy have a net negative impact on the soil organic carbon and on the environment through the release of CO_2. For instance, in Ghana, Parker et al. [22] documented a 21% reduction in soil organic carbon because of biomass burning that resulted in the release of 1.4 t CO_2ha^{-1} to the atmosphere. A study in southern Asia also demonstrated that burning of 5–7tha^{-1} of rice residues causes air pollution through releasing 13 t CO_2ha^{-1} [23].

In the tropical agricultural systems, the removal of crop residues for or by livestock, either through grazing or cut and carry, is a common practice, which conflicts the use of crop residues and cover crops for soil improvement [24]. Smith et al. [5] showed that it was more likely to observe an effect of straw removal on SOC: (i) in the less fertile soils, (ii) when greater quantities of residues were removed, and (iii) over longer periods. In line with this negative effect of residue removal on soil carbon, Nandwa [25] demonstrated that residue removal for off-farm use should consider only amounts that can be harvested without decreasing SOC levels. However, livestock are an important part of production in mixed farming systems and in the absence of alternative feed sources; farmers are usually unwilling to abandon this critically important one [25].

There is a consensus among the literature that most of the tropical agricultural systems lead to the depletion of organic matter due to the long-term extractive agricultural practices and reduced organic inputs, which consequently make most of the agricultural systems unsustainable. Due to this, there is an urgent need to improve the management of organic inputs and soil organic matter dynamics in tropical land use systems. One desirable goal is the ability to be able to manipulate soil organic matter dynamics via management practices so as to promote soil conservation, to ensure the sustainable productivity of agroecosystems, and to increase the capacity of tropical soils to act as a sink for, rather than a source of, atmospheric carbon.

LAND AND CROP MANAGEMENT PRACTICES FOR SOIL C SEQUESTRATION

In the last two to three decades, several land and crop management practices have been advocated to restore soil organic carbon and reduce net emissions of CO_2 from the agricultural systems in the tropics [5, 26, 27]. Among others, practices that restore soil organic carbon and reduce net emissions of CO_2 include crop rotation, avoiding use of bare fallow, conservation tillage, and management of organic inputs such as manure and crop residues, restoration of degraded agricultural lands, water management, and agroforestry (e.g., 28–34]).

Studies demonstrated that smallholder farmers can reduce greenhouse gas emissions and maintain carbon stocks in soil and vegetation at relatively low cost by implementing crop and land management practices (e.g., [27, 35–37]). However, a review by Giller et al. [38] and Sanchez, 2000 [39], identified a number of constraints that include a low degree of mechanization within the smallholder system, lack of appropriate implements, problem of weed control under no-till system, and lack of appropriate technical information that hinders large-scale adoption of the practices by the smallholder farmers. Woodfine [27] added that a key bottleneck to realizing the adoption of many mitigation practices is the availability of financing to catalyze initial change. Operationally, improved crop and land management practices may require more manual labor than conventional agricultural practices [40]. Optimizing these advantages and disadvantages can be a complex task which is in itself a disadvantage where there is a scarcity of trained personnel and extension workers to provide information and advice to farmers.

Furthermore, the temporal pattern of influence in mitigating the increase in CO_2 varies among practices and, in most cases, CO_2 emissions reduction resulted from the advocated practices are temporary [5]. For example, a study in Kenya documented that the residual effect of manure applied for four years only lasted another seven or eight years when assessed by yield, SOC, and Olsen P [41]. Effect of no-till practices are also easily reversed and lead to the release of CO_2 to the atmosphere as soon as the system started to be disturbed.

In sum, under present circumstances, recommended crop and land management practices are inappropriate for the vast majority of resource constrained smallholder farmers and farming systems [38]. However, this does not mean that mitigation practices advocated in the last two to three decades could not be one option that can offer substantial benefits for smallholder farmers in the tropics who are not constrained by resources and in certain locations where political, economical, and institutional frame conditions are relatively efficient. Identification of the situations when mitigation practices can offer major benefits is a challenge that demands active research [38].

BIOCHAR AS A CLIMATE CHANGE MITIGATION OPTION

Biochar is a charcoal produced under high temperatures (300 to 500°C) through the process of pyrolysis using crop residues, animal manure, or any type of organic waste material [42]. The two main methods of pyrolysis are "fast" pyrolysis and "slow" pyrolysis. Fast pyrolysis yields 60% bio-oil, 20% biochar, and 20% syngas and can be done in seconds, whereas slow pyrolysis can be optimized to produce substantially more char (~50%), but takes on the order of hours to complete [43]. Depending on the feedstock, biochar may look similar to potting soil or to a charred substance. The combined production and use of biochar are considered a carbon-negative process, meaning that it removes carbon from the atmosphere [42, 44]. Studies suggest that biochar sequester approximately 50–80% of the carbon available within the biomass feedstock being pyrolyzed depending upon the feedstock type [45, 46].

CAN WE PRODUCE BIOCHAR USING LOCALLY AVAILABLE TECHNOLOGIES?

Biochar can be produced using locally made technologies, which can be affordable to the local farmers and easily adopted and used. One of such easy technologies is the use of biochar chamber made of stainless steel (Figure 1). This kind of technology only costs about $ 70USper chamber (based on the amount of money invested to produce a chamber at Laos PDR), and the system operation and maintenance are quite easy and can be managed by the smallholder farmers in the tropics. We have also measured that it has the capacity to produce 83.3 (±4.2) kg of biochar from a rice husk with conversion efficiency of 48.1 (±2.1) % per 14.5 (±1.0) hours of burning. Other methods used to produce biochar in small quantity for use by the small-scale farmers that are described in http://www.biochar.info/ include carbon zero experimental biochar kiln, simple two-barrel biochar retort, and simple two barrel biochar retort with afterburner.

Biochar chamber made of stainless steel

(a)

The chamber covered with rice husk and burning

(b)

Biochar produced after 14-18 hours of burning

(c)

Figure 1: The process of biochar production using biochar chamber made of stainless steel from a rice husk (photo by Wolde Mekuria).

In addition, in The Netherlands, a "Twin-retort" carbonization process has been developed to address charcoal production efficiency and emission problems [47]. The traditional charcoal production systems used in the past such as charcoal production in open pits, earthen kilns, and traditional charcoal mounds, as carried out in rural areas, are inefficient. In most cases, weight efficiency of the traditional charcoal production systems carried out in rural areas ranged from 10 to 15% indicating that seven to ten kilograms of wood are required to produce one kilogram of charcoal [47]. Reumerman and Frederiks [47] documented that the efficiency of"Twin-retort" carbonization process is more than double compared to the tradition charcoaling processes. This indicates the possibility to reduce emissions with at least a factor of two.

We argue that the possibility to produce biochar using simple and locally available technologies speeds up the adoption of biochar production systems and its use as a climate change mitigation measure and improving agricultural productivity provided that obstacles that may halt rapid adoption of biochar production systems include technology costs, system operation, and maintenance [42]. Although biochar research and development are in their early stage, interest in biochar as a tool to mitigate the increase in CO_2 and improve agricultural productivity is growing at a rapid pace across the tropics.

BIOCHAR, SOIL C, SOIL FERTILITY, AND PRODUCTIVITY

Soil amendment with biochar has been proposed as a means to sequester C (Table 1) and improve soil fertility. Application of charcoal to soils is hypothesized to increase bioavailable water, build soil organic matter, enhance nutrient cycling, lower bulk density, act as a liming agent, and reduce leaching of pesticides and nutrients to surface and ground water [48–51]. Leach et al. [52] also documented that application of biochar to the soil enabling increases in agricultural productivity without, or with much reduced, applications of inorganic fertilizer. Furthermore, Harley [53] indicted that biochar is a promising amendment for ameliorating drastically disturbed soils due to its microchemical, nutrient, and biological properties (Table 2). Biochar-based strategies are thus being seen to offer valuable routes to building sustainable agricultural futures, particularly for resource poor farmers for whom soil fertility and water availability are seen as key constraints on crop production and food security [52].

Table 1: Effects of biochar sourced from different biomass on soil C

Country	Soil type	Treatment	Application rate	Changes in soil C*	Source	Remark
Philippines	Gleysols	Rice husk biochar	41.3tha−1	12.9gkg−1	[54]	After 3 years
Philippines	Nitosols	Rice husk biochar	41.3tha−1	12.4gkg−1	[54]	After 3 years
Thailand	Acrisols	Rice husk biochar	41.3tha−1	0.51gkg−1	[54]	After 3 years
Ethiopia	Nitosols	Maize stalk biochar	5tha−1	0.71%	[55]	Incubation trial
Ethiopia	Nitosols	Maize stalk	10tha−1	0.77%	[55]	Incubation trial
South Africa	Acidic sandy soils	Pinewood sawmill biochar	10tha−1	8.11%	[56]	Pot trial
India	Vertic ustropept	Prosopis biochar	5% of the incubated soil	4.5gkg−1	[57]	After 90 days of incubation

Kenya	Ferrasol	Acacia tree biochar	50tha−1	0.7%	[58]	Greenhouse experiment

Changes in soil C refers to the increase in C due to addition of biochar against the control plots.

Table 2: Role of biochar in ameliorating drastically disturbed soils [53]

Limiting factor	Variable	Problem	Short-term treatment	Long-term treatment	Role of biochar
Physical	Soil structure	Soil too compact	Rip or scarify	Vegetation	Decreased soil bulk density, increased infiltration, and decreased erodibility
	Soil erosion	High erodibility	Mulch	Regrade vegetation	
	Soil moisture	Too wet	Drain	Wetland construction	
		Too dry	Organic mulch	Tolerant species	Increased water retention due to surface area and charge characteristics
Nutritional	Macronutrients	Nitrogen deficiency Other deficiencies	Fertilizer Fertilizer	Nitrogen fixing plants, for example, leguminous trees or shrubs Fertilizer, amendments, tolerant species	Yield increases Slow nutrient release Soil organic matter stabilization Retention of released nutrients Increased microbial activity Habitat for mychorrhizal fungi hyphae

Toxicity	pH	Acid soils (<4.5)	Lime	Tolerant species	Designed for alkaline surface charge
		Alkaline soils (>7.8)	Pyritic waste, organic matter	Weathering, tolerant species	High CEC for Na retention
	Heavy metals	High concentration	Organic matter, tolerant cultivar	Inert covering, tolerant species	High surface area and cation exchange capacity allows for metal retention
	Salinity	EC > 4ds/m pH < 8.5, SAR < 13	Gypsum, irrigation	Weathering, tolerant species	Mixed with gypsum to reduce soil structural issues
	Sodicity	EC < 4ds/m, pH > 8.5, SAR ≥ 13	Gypsum, irrigation	Weathering, tolerant species	Nutritional values as described High CEC for Na retention

The extent of the effect of biochar on crop productivity and soil carbon sequestration is, however, variable due mainly to the different biophysical interactions and processes that occur when biochar is applied to soil, which are not yet fully understood [59]. For instance, in nitrogen limited soils, application of high rates of biochar may affect growth negatively due to immobilization effect [46]. Moreover, feedstock and pyrolysis conditions (temperature, holding time, etc.) may affect both stability and nutrient content and availability of biochar [59–61]. Given how inconsistent biochar impacts on yields and soil carbon sequestration are and how little is known about their longer-term impacts, farmers who are to use biochar on their fields are taking considerable risks such as a possible reductions in crop yield during the early cropping seasons.

Thus, we argue that care should be taken on the amount and type of biochar added to the soil for restoring degraded soils. In addition, it is crucial to detect the consequent soil organic carbon accumulation and increase in crop yields under different soil and climatic conditions. Long-term studies on biochar in field trials are also required to better understand biochar effects and to investigate its behavior in soils, thereby reducing the associated risks.

POTENTIAL OF BIOCHAR TO MITIGATE THE INCREASE IN CO$_2$

Studies have shown that cover crops, mulches, compost, or manure can be effective in enhancing soil organic carbon pool and agricultural productivity in the tropics (e.g., [29, 34, 62]). The benefits of such amendments are, however, often short-lived, especially in the tropics, since decomposition rates are high, and the added organic matters are usually mineralized to CO$_2$ within only a few cropping seasons. Organic amendments therefore have to be applied intermittently to sustain soil productivity. In case of agricultural lands converted to no-tillage systems, stored carbon can be released once we convert no-tillage back to conventional tillage. Therefore, carbon sequestered by these crop and soil management practices is generally considered only temporarily sequestered from the atmosphere and associated with a high risk of rapid or large-scale leakage [44].

Management of black carbon (C), increasingly referred to as biochar, may overcome some of those limitations and provide an additional soil management option. Once biochar is incorporated into soil, it is difficult to imagine any incident or change in practice that would cause a sudden loss of stored carbon indicating that biochar is a lower-risk strategy than other sequestration options [44]. Thus, biochar could be a potentially a powerful tool for mitigating anthropogenic climate change as the carbon in biochar, it is claimed, resists degradation and can sequester carbon in soils for hundreds to thousands of years [26, 44, 54, 63, 64]. The half-life of C in soil charcoal is in excess of 1000yr [65]. Laird et al. [49] presented an interesting graph that compares the stability of organic input added as residue biomass and biochar (Figure 2).

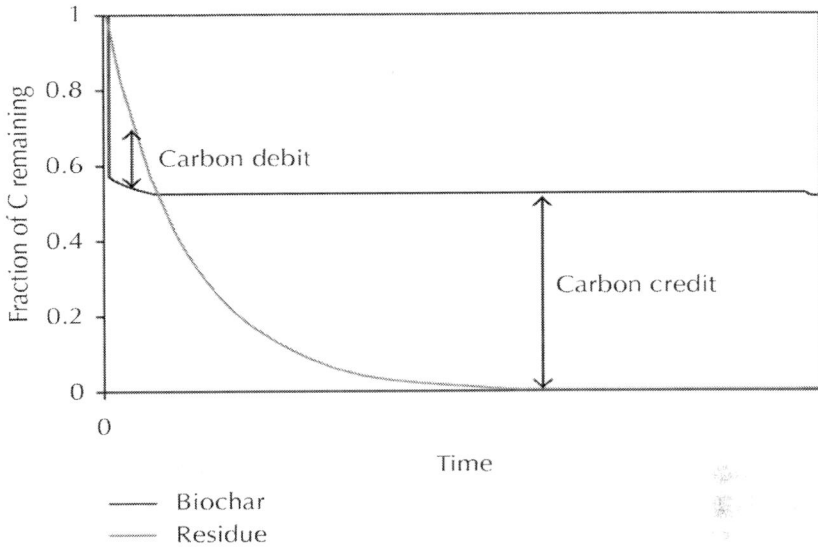

Figure 2: Impact of biomass pyrolysis with soil application of biochar on the amount of original biomass C remaining in the soil relative to the amount of C remaining in the soil if the same biomass is returned to the soil as a biological residue (source: Laird et al., 2009 [49]).

According to Laird et al. [49] (Figure 2) "For the Biochar example, about 40% of the C is lost at time 0 when the biomass is pyrolyzed, 10% of the total C is lost to mineralization over a few months, and the remaining 50% of the total C is stable for millennia. For the Residue example, the half-life of the residue C is assumed to be 6 months and 99% of the C is lost to mineralization after 4 years. The biochar scenario results in a C debit for the first 6 months and a C credit thereafter relative to the residue scenario". This indicates that biochar is a highly stable form of carbon and as such has the potential to form an effective C sink, therefore sequestering atmospheric CO_2 [59].

Current analyses suggest that there is global potential for annual sequestration of atmospheric CO_2 at the billion-tone scale (109 t yr^{-1}) within 30 years [59]. Woolf et al. [66] also indicated that annual net emissions of carbon dioxide (CO_2) could be reduced by a maximum of 1.8Pg CO_2-C equivalent (CO_2-C e) per year (12% of current anthropogenic CO_2-C e emissions; 1Pg = 1Gt) and total net emissions over the course of a century by 130Pg CO_2-C e, without endangering food security, habitat, or soil conservation. In addition, Gaunt and Lehmann [67] documented that emission reductions can be 12–84% greater if biochar is put back into the soil instead of being burned to offset fossil fuel use. Furthermore, the application of biochar once in every ten years to the estimated 15×10^9ha of cropland worldwide would result in a CO_2-equivalent gain of 0.65GtCyr^{-1}[68, 69]. These in turn indicate that biochar sequestration could be one option to change bioenergy into a carbon-negative industry [44

OPPORTUNITIES AND CONSTRAINTS FOR MITIGATING CARBON EMISSIONS IN TROPICAL AGRICULTURAL PRODUCTION SYSTEMS

In the tropics, soil organic carbon in the agricultural landscape has been depleted through the long-term use of extractive farming practices [18, 70, 71]. Most agricultural soils have lost 30 to 40tCha^{-1}, and their current reserves of soil organic carbon are much lower than their potential capacity indicating that there is great technical potential to increase soil carbon in agricultural soils and reduce greenhouse gas emissions. According to Lal [71], most soils have a technical or maximum sink capacity from 20 to 50tCha^{-1} that can be sequestered over a 20-to-50-year period. The greatest potential for sequestration is in the soils of those regions that have lost the most soil carbon. These are the regions where soils are severely degraded and have been used with extractive farming practices for a long time. Among developing countries, these regions include Sub-Saharan Africa, South and Central Asia, the Caribbean, Central America, and the Andean regions [71]. Among others, converting degraded soils into restorative land and adopting practices such as no-till, organic C input management such as additions of manure, compost, biochar, and agroforestry are practices that can increase the soil carbon pool in the tropics [71].

The literature reveals that in the last two to three decades that enabling and encouraging broader adoption of mitigation options were advocated through market-based mechanisms. This could create the catalyst necessary to elevate agriculture's role as a key part of a global approach to mitigating climate change. However, soil C sequestration through the advocated crop and soil management practices currently does not fit under emission trading (ET), the clean development mechanism (CDM), or joint implementation (JI), and neither Article 3.3 nor 3.4 of the Kyoto Protocol specifically include soil carbon as an option [72]. According to Lehmann [44], some of the reasons why soil C sequestration through crop and soil management practices is not allowed into trading markets under current agreements include (1) the net withdrawal of CO_2 through the advocated crop and soil management practices is usually short-lived, (2) accountability of the process of C sequestration is not straightforward, and monitoring and certifying the changes in C stocks are difficult and costly, and (3) the processes of C sequestration also associated with rapid or large-scale leakages. Sohi and Shackley [26] also pointed out that decomposition rate may increase with climate change making soil carbon stores vulnerable to "feedback." Only a small proportion of added organic matter stabilized for longer time, and accumulation rate diminishes with time resulting in inefficient use of organic resource after equilibration.

Yet, considering soil carbon as a commodity and creating another income stream for resource poor and small size land holders are essential prerequisite to widespread adoption of recommended management practices in the developing countries where the problems of food insecurity and soil/environmental degradation are extremely severe. Realizing these incomes would necessitate substantially greater policy support and investment in sustainable land uses than is currently the case [73]. Furthermore, while soil and crop management practices that enhance soil carbon pool in general clearly offer economic and ecological advantages, the development of robust systems compliant with stakeholder needs and requirements is constrained by our limited understanding of the tradeoffs between subsistence requirements, acceptable risks, and the costs involved [74].

CAN WE INTEGRATE BIOCHAR INTO TRADING MARKETS UNDER CURRENT AGREEMENTS?

Funding from carbon trading is argued to be essential to finance the research and development necessary to discover and exploit the full potential of crop and soil management practices to contribute to climate change mitigation and to enable wider adoption of the practices to sequester carbon at globally significant levels [52]. When it comes to including biochar in emission-trading schemes, the issues of permanence, land tenure, leakage, and additionality are less significant for biochar projects than for projects that sequester C in biomass or soil through management of plant productivity [44]. This is because biochar carbon sequestration might avoid difficulties such as accurate monitoring of soil carbon, which are the main barriers to inclusion of agricultural soil management in emissions trading [44, 75]. In addition, no complex predictive models or analytical tools are required, as is the case with other soil sequestration approaches. The source of biochar additions can easily be identified by soil analyses, if desired for verification under carbon-trading schemes.

We argue that there is a great potential to allow the associated emission reductions through using biochar into trading markets under current agreements, because emission reduction units obtained due to the use of biochar can easily be accounted, monitored, and verified. In addition, climate change is real that it will intensify, and there is an urgency not only to identify but also to implement solutions. Biochar sequestration does not require a fundamental scientific advance, and the underlying production technology is robust and simple, making it appropriate for many regions of the world [44]. Furthermore, the possibility to produce biochar using locally available technologies (Figure 1) even makes it more appropriate for the resource poor smallholder farmers living throughout the tropics.

CONSIDERATIONS IN UPSCALING BIOCHAR

Recognizing that biochar technology is in its early stages of development, there are many concerns about the applicability of the technology in the tropics. Three issues are feedstock availability, biochar handling, and biochar system deployment. To date, feedstock for biochar has consisted mostly of plant and crop residues, a primary source of energy and livestock feed for the smallholder farmers in the tropics. Thus, there is still sustainability concerns related to supplying feedstock for large-scale biochar production. The ideal time to apply biochar and how to ensure that it remains in place once applied and does not cause a risk to human health or degrade air quality is also a concern. Furthermore, developing a "one size fits all" biochar system would also be a challenge as biochar systems are designed on the feedstock to be decomposed and the energy needs of an operation [42].

The literature indicates that biochar can be effective in improving soil organic C, nutrient cycling, and crop yield (e.g., [49]). However, biochar production involves removal of crop residues from agricultural lands and would increase risk of accelerated erosion. Thus, determination of sustainable crop residue removal rates and implementation of additional conservation practices such as contour cropping, conservation tillage, and cover crops in agricultural lands are crucial. Furthermore, competition with food production and induced land use change would diminish the carbon sequestration potential even for a strategy as promising as biochar [75]. As biochar carbon sequestration depends on revenues from carbon trading, it is important to ensure that large-scale biochar application on agricultural lands will not lead to depleting the terrestrial carbon stock as it reduces the economic viability of biochar.

We argue that it is important to consider issues such as feedstock availability while promoting biochar as climate change mitigation option in the tropics as the farming system in the tropics is dominated by mixed crop-livestock production systems. Under such system there is always a competition in the use of crop residues for soil amendments or for livestock feed. However, this conflicting issue can be resolved by arranging alternative feedstocks to feed the livestock.

REFERENCES

1. IPCC, "Climate change 2007: mitigation," in Contribution of Working Group III to the Fourth Assessment Report of the Intergovernmental Panel on Climate Change, B. Metz, O. R. Davidson, P. R. Bosch, R. Dave, and L. A. Meyer, Eds., p. 30, Cambridge University Press, Cambridge, UK, 2007.

2. M. J. Salinger, M. V. K. Sivakumar, and R. Motha, "Reducing vulnerability of agriculture and forestry to climate variability and change: workshop summary and recommendations," Climatic Change, vol. 70, no. 1-2, pp. 341–362, 2005.

3. C. E. P. Cerri, G. Sparovek, M. Bernoux, W. E. Easterling, J. M. Melillo, and C. C. Cerri, "Tropical agriculture and global warming: impacts and mitigation options," Scientia Agricola, vol. 64, no. 1, pp. 83–99, 2007.

4. S. Jaiaree, A. Chidthaisong, and N. Tangtham, "Soil carbon dynamics and net carbon dioxide fluxes in tropical forest and corn plantation system," in Proceedings of the 2nd Joint International Conference on Sustainable Energy and Environment (SEE '06), Bangkok, Thailand, November 2006.

5. P. Smith, D. Martino, Z. Cai et al., "Greenhouse gas mitigation in agriculture," Philosophical Transactions of the Royal Society B, vol. 363, no. 1492, pp. 789–813, 2008.

6. S. M. Ogle, F. J. Breidt, and K. Paustian, "Agricultural management impacts on soil organic carbon storage under moist and dry climatic conditions of temperate and tropical regions," Biogeochemistry, vol. 72, no. 1, pp. 87–121, 2005.

7. J.M.-F. Johnson, A. J. Franzluebbers, S. L. Weyers, and D. C. Reicosky, "Agricultural opportunities to mitigate greenhouse gas emissions," Environmental Pollution, vol. 150, no. 1, pp. 107–124, 2007.

8. FAO, Enabling Agriculture to Contribute to Climate Change Mitigation, The Food and Agriculture Organization of the United Nations, Rome, Italy, 2010.

9. FAO, "Agriculture and environmental challenges of the twenty-first century: a strategic approach for FAO," Tech. Rep. COAG/2009/3, 11, FAO, Rome, Italy, 2009.

10. A. S. Grandy and G. P. Robertson, "Land-use intensity effects on soil organic carbon accumulation rates and mechanisms," Ecosystems, vol. 10, no. 1, pp. 58–73, 2007.

11. E. Boserup, The Conditions of Agricultural Growth: The Economics of Agrarian Change under Population Pressure, G. Allen and Unwin, London, UK, 1965.

12. C. L. A. Asadu, F. I. Nweke, and A. A. Enete, "Soil properties and intensification of traditional farming systems in Sub Saharan Africa (SSA)," Journal of Tropical Agriculture, Food, Environment and Extension, vol. 7, pp. 186–192, 2008.

13. K. Vielhauer, T. D. A. Sa, and M. Denich, "Modification of a traditional crop-fallow system towards ecologically and economically sound options in the eastern Amazon," in Proceedings of the German-Brazilian Workshop on Neotropical Ecosystems—Achievements and Prospects of Cooperative Research, Hamburg, Germany, September 2000.

14. S. G. Perz, "Household demographic factors as life cycle determinants of land use in the Amazon,"Population Research and Policy Review, vol. 20, no. 3, pp. 159–186, 2001.

15. M. J. Kipsat, P. M. Anangweso, M. K. Korir, A. K. Serem, H. K. Maritim, and B. J. Kanule, "Factors affecting farmers'decisions to abandon soil conservation once external support cease: the case of Kericho District, Kenya," African Crop Science Conference Proceeding, vol. 8, pp. 1377–1381, 2007.

16. S. Holden and H. Yohannes, "Land redistribution, tenure insecurity, and intensity of production: a study of farm households in Southern Ethiopia," Land Economics, vol. 78, no. 4, pp. 573–590, 2002.

17. P. Ebanyat, N. de Ridder, A. de Jager, R. J. Delve, M. A. Bekunda, and K. E. Giller, "Drivers of land use change and household determinants of sustainability in smallholder farming systems of Eastern Uganda," Population and Environment, vol. 31, no. 6, pp. 474–506, 2010.

18. R. Lal, "Soil carbon sequestration impacts on global climate change and food security," Science, vol. 304, no. 5677, pp. 1623–1627, 2004.

19. A. Bationo, J. Kihara, B. Vanlauwe, B. Waswa, and J. Kimetu, "Soil organic carbon dynamics, functions and management in

West African agro-ecosystems," Agricultural Systems, vol. 94, no. 1, pp. 13–25, 2007.

20. P.-A. Jacinthe, R. Lal, and J. M. Kimble, "Carbon dioxide evolution in runoff from simulated rainfall on long-term no-till and plowed soils in southwestern Ohio," Soil and Tillage Research, vol. 66, no. 1, pp. 23–33, 2002.

21. Y. Abera and T. Belachew, "Effects of landuse on soil organic carbon and nitrogen in soils of bale, Southeastern Ethiopia," Tropical and Subtropical Agroecosystems, vol. 14, no. 1, pp. 229–235, 2011.

22. B. Q. Parker, B. A. Osei, F. A. Armah, and D. O. Yawson, "Impact of biomass burning on soil organic carbon and the release of carbon dioxide into the atmosphere in the coastal savanna ecosystem of Ghana," Journal of Renewable and Sustainable Energy, vol. 2, no. 3, Article ID 033106, 2010.

23. K. G. Mandal, A. K. Misra, K. M. Hati, K. K. Bandyopadhyay, P. K. Ghosh, and M. Mohanty, "Rice residue-management options and effects on soil properties and crop productivity," Food, Agriculture & Environment, vol. 2, pp. 224–231, 2004.

24. V. Smil, "Crop residues: agriculture's largest harvest," BioScience, vol. 49, no. 4, pp. 299–308, 1999.

25. S. M. Nandwa, "Soil organic carbon (SOC) management for sustainable productivity of cropping and agro-forestry systems in Eastern and Southern Africa," Nutrient Cycling in Agroecosystems, vol. 61, no. 1-2, pp. 143–158, 2001.

26. S. P. Sohi and S. Shackley, "Biochar: carbon sequestration potential," December 2009, Copenhagen, Denmark.

27. A. Woodfine, "Using sustainable land management practices to adapt to and mitigate climate change in sub-saharan Africa," 2009, Resource guide version 1. 0. TERR AFRICA, http://www.terrafrica.org/.

28. P. K. R. Nair, V. D. Nair, E. F. Gama-Rodrigues, et al., "Soil carbon in agro forestry systems: an unexplored treasure," Nature Proceedings. In press.

29. K. Banger, G. S. Toor, A. Biswas, S. S. Sidhu, and K. Sudhir, "Soil organic carbon fractions after 16-years of applications of fertilizers and organic manure in a Typic Rhodalfs in semiarid

tropics," Nutrient Cycling in Agroecosystems, vol. 86, no. 3, pp. 391–399, 2010.

30. L. Batlle-Bayer, N. H. Batjes, and P. S. Bindraban, "Changes in organic carbon stocks upon land use conversion in the Brazilian Cerrado: a review," Agriculture, Ecosystems and Environment, vol. 137, no. 1-2, pp. 47–58, 2010.

31. J. Fallahzade and M. A. Hajabbasi, "The effects of irrigation and cultivation on the quality of desert soil in central Iran," Land Degradation and Development, vol. 23, no. 1, pp. 53–61, 2012.

32. M. Shafi, J. Bakht, A. Attaullah, and M. A. Khan, "Effect of crop sequence and crop residues on soil C, soil N and yield of maize," Pakistan Journal of Botany, vol. 42, no. 3, pp. 1651–1664, 2010.

33. Q. Wang, Y. Li, and A. Alva, "Cropping systems to improve carbon sequestration for mitigation of climate change," Journal of Environmental Protection, vol. 1, pp. 207–215, 2010.

34. S. A. Bangroo, N. K. Kirmani, T. Ali, M. A. Wani, M. A. Bhat, and M. I. Bhat, "Adapting agriculture for enhancing ecoefficiency through soil carbon sequestration in agro-ecosystem," Research Journal of Agricultural Sciences, vol. 2, pp. 164–169, 2011.

35. D. N. Pandey, "Carbon sequestration in agroforestry systems," Climate Policy, vol. 2, no. 4, pp. 367–377, 2002.

36. A. A. Kimaro, V. R. Timmer, A. G. Mugasha, S. A. O. Chamshama, and D. A. Kimaro, "Nutrient use efficiency and biomass production of tree species for rotational woodlot systems in semi-arid Morogoro, Tanzania," Agroforestry Systems, vol. 71, no. 3, pp. 175–184, 2007.

37. P. K. R. Nair, "The coming of age of agroforestry," Journal of the Science of Food and Agriculture, vol. 87, no. 9, pp. 1613–1619, 2007.

38. K. E. Giller, E. Witter, M. Corbeels, and P. Tittonell, "Conservation agriculture and smallholder farming in Africa: the heretics' view," Field Crops Research, vol. 114, no. 1, pp. 23–34, 2009.

39. P. A. Sanchez, "Linking climate change research with food security and poverty reduction in the tropics," Agriculture, Ecosystems and Environment, vol. 82, no. 1–3, pp. 371–383, 2000.

40. D. Suprayogo, K. Hairiah, M. V. Noordwijk, and G. Cadisch, "Agroforestry interactions in rainfed agriculture: can hedgerow

intercropping systems sustain crop yield on an ultisol in lampung (Indonesia)?" Agrivita, vol. 32, no. 3, 2010.

41. F. M. Kihanda, G. P. Warren, and A. N. Micheni, "Effect of manure application on crop yield and soil chemical properties in a long-term field trial of semi-arid Kenya," Nutrient Cycling in Agroecosystems, vol. 76, no. 2-3, pp. 341–354, 2006.

42. K. Bracmort, "Biochar: examination of an emerging concept to mitigate climate change," CRS Report for Congress, United States Congressional Research Service, 2010.

43. I. F. Odesola and T. A. Owoseni, "Development of local technology for a small-scale biochar production processes from agricultural wastes," Journal of Emerging Trends in Engineering and Applied Sciences, vol. 1, no. 2, pp. 205–208, 2010.

44. J. Lehmann, "A handful of carbon," Nature, vol. 447, no. 7141, pp. 143–144, 2007.

45. J. Lehmann, J. P. da Silva Jr., M. Rondon et al., "Slash-and-char—a feasible alternative for soil fertility management in the central Amazon?" in Proceedings of the 17th World Congress of Soil Science, CD-ROM Paper no. 449, pp. 1–12, Bangkok, Thailand, 2002.

46. J. Lehmann, J. Gaunt, and M. Rondon, "Bio-char sequestration in terrestrial ecosystems—a review,"Mitigation and Adaptation Strategies for Global Change, vol. 11, no. 2, pp. 403–427, 2006.

47. P. J. Reumerman and B. Frederiks, "Charcoal production with reduced emissions," in Proceedings of the 12th European Conference on Biomass for Energy, Industry and Climate Protection, Amsterdam, The Netherlands, 2002.

48. D. A. Laird, "The charcoal vision: a win-win-win scenario for simultaneously producing bioenergy, permanently sequestering carbon, while improving soil and water quality," Agronomy Journal, vol. 100, no. 1, pp. 178–181, 2008.

49. D. A. Laird, R. C. Brown, J. E. Amonette, and J. Lehmann, "Review of the pyrolysis platform for coproducing bio-oil and biochar," Biofuels, Bioproducts and Biorefining, vol. 3, no. 5, pp. 547–562, 2009.

50. J. M. Novak, W. J. Busscher, D. L. Laird, M. Ahmedna, D. W. Watts, and M. A. S. Niandou, "Impact of biochar amendment on

fertility of a southeastern coastal plain soil," Soil Science, vol. 174, no. 2, pp. 105–112, 2009.

51. P. Brookes, L. Yu, M. Durenkam, and Q. Lin, "Effects of biochar on soil chemical and biological properties in high and low pH soils," in Proceedings of the International Symposium on Environmental Behavior and Effects of Biomass-Derived Charcoal, China Agricultural University, Beijing, China, October 2010.

52. M. Leach, J. Fairhead, J. Fraser, and E. Lehner, "Biocharred pathways to sustainability? Triple wins, livelihoods and the politics of technological promise," STEPS Working Paper 41, STEPS Centre, Brighton, UK, 2010.

53. A. Harley, "Biochar for reclamation," in The Role of Biochar in the Carbon Dynamics in Drastically Disturbed Soils. US-Focused Biochar Report, US Biochar Initiative, 2010.

54. S. M. Haefele, Y. Konboon, W. Wongboon et al., "Effects and fate of biochar from rice residues in rice-based systems," Field Crops Research, vol. 121, no. 3, pp. 430–440, 2011.

55. A. Nigussie, E. Kissi, M. Misganaw, and G. Ambaw, "Effect of biochar application on soil properties and nutrient uptake of lettuces (Lactuca sativa) grown in chromium polluted soils," American-Eurasian Journal of Agricultural & Environmental Science, vol. 12, pp. 369–376, 2012.

56. A. Hardie and A. Botha, Biochar Amendment of Infertile Western Cape Sandy Soil: Implications for Food Security, Stellenbosch University, Stellenbosch, South Africa.

57. S. Shenbagavalli and S. Mahimairaja, "Characterization and effect of biochar on nitrogen and carbon dynamics in soil," International Journal of Advanced Biological Research, vol. 2, pp. 249–255, 2012.

58. C. Söderberg, Effects of biochar amendment in soils from Kisumu, Kenya [Degree project in Biology, SLU], Swedish University of Agricultural Sciences, Faculty of Natural Resources and Agricultural Sciences, Department of Soil and Environment, Uppsala, Sweden, 2013.

59. S. Sohi, E. Lopez-Capel, E. Krull, and R. Bol, "Biochar, climate change and soil: a review to guide future research," CSIRO Land and Water Science Report 05/09, CSIRO, Highett, Australia, 2009.

60.   J. W. Gaskin, C. Steiner, K. Harris, K. C. Das, and B. Bibens, "Effect of low-temperature pyrolysis conditions on biochar for agricultural use," Transactions of the ASABE, vol. 51, no. 6, pp. 2061–2069, 2008.

61.   J. M. Novak, I. Lima, B. Xing, et al., "Characterization of designer biochar produced at different temperatures and their effects on a loamy sand," Annals of Environmental Science, vol. 3, pp. 195–206, 2009.

62.   B. Barthès, A. Azontonde, E. Blanchart et al., "Effect of a legume cover crop (Mucuna pruriens var. utilis) on soil carbon in an Ultisol under maize cultivation in southern Benin," Soil Use and Management, vol. 20, pp. 231–239, 2004.

63.   C. Cheng, J. Lehmann, J. E. Thies, S. D. Burton, and M. H. Engelhard, "Oxidation of black carbon by biotic and abiotic processes," Organic Geochemistry, vol. 37, no. 11, pp. 1477–1488, 2008.

64.   J. Lehmann, "Biological carbon sequestration must and can be a win-win approach," Climatic Change, vol. 97, no. 3, pp. 459–463, 2009.

65.   B. Glaser, J. Lehmann, C. Steiner, T. Nehls, M. Yousaf, and W. Zech, "Potential of pyrolyzed organic matter in soil amelioration," in Proceedings of the 12th ISCO Conference, Beijing, China, 2002.

66.   D. Woolf, J. E. Amonette, F. A. Street-Perrott, J. Lehmann, and S. Joseph, "Sustainable biochar to mitigate global climate change," Nature Communications, vol. 1, no. 5, article 56, 2010.

67.   J. Gaunt and J. Lehmann, "Prospects for carbon trading based in the reductions of greenhouse gas emissions arising from the use of biochar," in Proceedings of the International Agrichar Initiative Conference (IAI '07), p. 20, Terrigal, Australia, April 2007.

68.   J. L. Gaunt and J. Lehmann, "Energy balance and emissions associated with biochar sequestration and pyrolysis bioenergy production," Environmental Science and Technology, vol. 42, no. 11, pp. 4152–4158, 2008.

69.   N. Ramankutty, A. T. Evan, C. Monfreda, and J. A. Foley, "Farming the planet: 1. Geographic distribution of global agricultural lands in the year 2000," Global Biogeochemical Cycles, vol. 22, no. 1, Article ID GB1003, 2008.

70. R. Lal, "Enhancing crop yields in the developing countries through restoration of the soil organic carbon pool in agricultural lands," Land Degradation and Development, vol. 17, no. 2, pp. 197–209, 2006.

71. R. Lal, "Challenges and opportunities in soil organic matter research," European Journal of Soil Science, vol. 60, no. 2, pp. 158–169, 2009.

72. FAO, Managing Soil Carbon to Mitigate Climate Change: A Sound Investment in Ecosystem Services, a Framework for Action, Food and Agriculture Organization of the United Nations Conservation Technology Information Center, Rome, Italy, 2008.

73. J. O. Niles, S. Brown, J. Pretty, A. S. Ball, and J. Fay, "Potential carbon mitigation and income in developing countries from changes in use and management of agricultural and forest lands," Philosophical Transactions of the Royal Society A, vol. 360, no. 1797, pp. 1621–1639, 2002.

74. K. P. C. Rao, L. V. Verchot, and J. Laarman, "Adaptation to climate change through sustainable management and development of agroforestry systems," ICRISAT, vol. 4, no. 1, pp. 1–30, 2007.

75. C. Steiner, "Biochar in agricultural and forestry applications," in Biochar from Agricultural and Forestry Residues—a Complimentary Use of "Waste" Biomass. Assessment of Biochar's Benefits for the United States of America, US Biochar Initiative, 2010.

Chapter 6

Tropical Legume Crop Rotation and Nitrogen Fertilizer Effects on Agronomic and Nitrogen Efficiency of Rice

Motior M. Rahman, Aminul M. Islam, Sofian M. Azirun, and Amru N. Boyce

Institute of Biological Sciences, Faculty of Science, University of Malaya, 50603 Kuala Lumpur, Malaysia

ABSTRACT

Bush bean, long bean, mung bean, and winged bean plants were grown with N fertilizer at rates of 0, 2, 4, and 6 g N m^{-2} preceding rice planting. Concurrently, rice was grown with N fertilizer at rates of 0, 4, 8, and 12 g N m^{-2}. No chemical fertilizer was used in the 2nd year of crop to estimate the nitrogen agronomic efficiency (NAE), nitrogen recovery efficiency (NRE), N uptake, and rice yield when legume crops were grown in rotation with rice. Rice after winged bean grown with N at the rate of $4\,g\,N\,m^{-2}$ achieved significantly higher NRE, NAE, and N uptake in both years. Rice after winged bean grown without N fertilizer produced 13–23% higher grain yield than rice after fallow rotation with $8\,g\,N\,m^{-2}$. The results revealed that rice after winged bean without fertilizer and rice after long bean with N fertilizer at the rate of $4\,g\,N\,m^{-2}$ can produce rice yield equivalent to that of rice after fallow with N fertilizer at rates of $8\,g\,N\,m^{-2}$. The NAE, NRE, and harvest index values for rice after winged bean or other legume crop rotation indicated a positive response for rice production without deteriorating soil fertility.

INTRODUCTION

Rice is the most widely consumed staple food and most commercially important crop for more than 3 billion people in the world's human population [1]. Nitrogen is quantitatively the most essential nutrient for plants [2] and a major constraint and contributing factor for low productivity and widespread food insecurity in most rice-based cropping systems in Asia [3, 4]. The intensive cultivation of cropping practices with high yielding rice varieties requires better soil and nutrient management [5]. Alternatively, long-term cropping can degrade soil fertility [6]. Soil texture and crop rotation practices can influence rates of N fertilizer application in rice crop [7, 8]. However, imbalanced rates and injudicious methods of fertilizer application can lead to poor N efficiency, N losses due to leaching, and other chemical and biological processes in soil [9–12], resulting in a series of environmental hazards and economic loses. Thus, it is obvious that poor nitrogen use efficiency (NUE) causes higher production costs and induces lower net returns for rice growers [13]. So, an efficient N utilization must be ensured for sustainable crop production for the benefit of environment and economic reasons [14, 15]. Optimum use of N fertilizer is a crucial step to improve NUE, while a positive relation between soil N supply and crop N demand is one of the key factors to appropriate N utilization by plants [16].

Soil N supply through biological nitrogen fixation (BNF) by associated microbial populations is one of the principal sources of N for rice production. However, the loss of soil N occurs continuously through removal of plant or harvesting of grain and chemical processes of the soil [12]. In addition, the indigenous soil N supply in wetland rice may decline with intensive rice cultivation unless it is restored by BNF [17]. Considering both environmental and economic perspectives, maintenance of native soil N resource and improvement of N output from plant sources are one of the desirable options to reduce the use of chemical fertilizer in rice cropping system [18]. Soil N loss may be minimized by using effective legume crops which can supply sufficient BNF input to enhance soil N by improved recycling of N through plant residues [19]. Thus, the combined indigenous soil N and N achieved through legume in BNF have the potential for N enrichment in soil which will increase NUE of crops and total N output in a lowland rice-based cropping system [20]. In addition to fixing atmospheric nitrogen, leguminous green manures play a significant role in conserving NO_3 [21, 22]. It is well recognized that the legume plant can add organic N and their residues contribute to the improvement of soil texture and microbial activity. However, rice growers believe that the accrued N benefits will vary among different legume systems [23].

In Malaysia, most of the rice growing areas are well established with irrigation systems and farmer's practicing double cropping systems with high yielding varieties of rice crop [5], which mainly depends on inorganic N fertilizer and other chemical fertilizers. The government has spent more than 3.0 billion US$ (RM 9.2 billion) to import 4.2 million tons of mineral fertilizers to sustain crop production in Malaysia [24]. Subsequently, its policy has focused more towards agroecological, healthier, and sustainable food production practices through an integrated approach of rice cultivation with crop rotation using vegetable, legume, and intercropping practices for sweet corn, maize, and organic farming to minimize the dependence on mineral fertilizers [25–27]. Inclusion of grain legumes or green manure legume crops in rotation with rice or corn can protect degradation of soil fertility [28–30], improve soil structure, water holding capacity [31], and result in greater productivity and higher income, while minimizing production risk and ensuring long-term sustainability, as well as ensure a greener environment [30, 31]. Incorporation of crop residues alters the soil environment that in turn influences the microbial population's activity in the soil and subsequent nutrient cycle [23] and will sustain rice productivity through replenishing soil organic matter [32]. The presence of soil organic matter is a key indicator of soil quality, which provides plant nutrients upon mineralization and eventually improves soil properties [33].

Legume crop residues contribute to organic N and after decomposition by soil microbes, through mineralization, add available N for the next crop [34], and ameliorate the nutrient status of the soil. In Malaysia, rice growers cultivate two rice crops per year and sometimes five crops in a two-year period, but crop rotation practices of rice with tropical grain legumes or green manure legume crops are not often used [27]. This has brought about soil fertility deterioration, which threatens the ecosystem through intensive application of inorganic chemical fertilizers. Thus, the appropriate management and efficient utilization of crop residues are important for the proper amendment of soil quality and crop productivity under a rice-based cropping system in the tropics [22]. Indeed, the use of grain legumes or cover crops has been proven to be commendable in terms of its positive effects. Bush bean, long bean, sprouted mung bean seed, and winged bean are traditional vegetables cultivated in marginal land and there are ample opportunities to adapt these crops in an upland rice-based crop rotation system. Presently, growers are concerned about the quality and as well sustainable use rather than the quantity of food production. Attention has been focused towards the use of legume crops to improve soil health for productivity of rice crop. The practice of using tropical legumes such as, bush bean, long bean, winged bean, or mung bean, alone or in combination with inorganic N fertilizers, offers promising scope as an N supplement to rice crop rotation systems. No systematic research has been carried out on the consequence of N in green manure legume and the productivity of legume crops and their effects on soil N dynamics and contributions to the yield and N uptake of the following rice crop in Malaysia. The combined effect of legume residues and indigenous nutrient supplies or fertilizer uptake and losses in rice-based systems is little understood. The present study was undertaken to assess the addition of legume residues to plant N uptake, NAE, and NRE and also the amount of fertilizer N essential for optimizing rice yield when legumes are enclosed in the system.

MATERIALS AND METHODS

Experimental Plan and Management

The experiments were carried out at the greenhouse, University of Malaya, Kuala Lumpur, Malaysia, during 2010 and 2011. The clay loam soil was collected at depth of 30 cm from rice field in Selangor (1° 28' 0" N, 103° 45' 0" E), Malaysia. No specific permits were required for the described studies and no specific permissions were required for these activities. In addition, the locations were not protected and the studies did not involve endangered or protected species. The soil used had the following chemical properties: pH 6.55 ± 0.20 (1 : 5 w/v water), CEC 15 (cmolc kg^{-1} soil), organic C 1.75 ± 0.48% (CHNS analyzer, model NA 1500), total N 0.18 ± 0.04%, NH4-N 6.37 ± 1.25 (mg 100−1 g soil), exchangeable CaO 171.0±21.15 (mg 100^{-1} g soil), exchangeable MgO 10.8±2.75 (mg 100−1 g soil), and exchangeable K$_2$O 14.9 ± 9.06 (mg 100^{-1} g soil). The soils were thoroughly mixed and unwanted inert materials were discarded through sieve (2 mm mesh) to produce homogenous soil composites. The experimental pots (height 46 cm × diameter 54 cm = surface area 1 m^2)were filled with soil up to about 36 cm height (height 36 cm × diameter 54 cm = surface area 0.84 m^2) of each pot. All data was converted into 1 m2 .The seeds of bush bean (Phaseolus vulgaris L.), long bean (Vigna unguiculata (L.) Verdc.),mung bean (Vigna radiata (L.) R. Wilczek), winged bean (Psophocarpus tetragonolobus (L.)D.C.), and corn (Zea mays L.) were sown in moistened soil to ensure germination. All legume crops were fertilized with N fertilizer (urea 46% N) at rates of 0, 2, 4, and 6 g N m^{-2} while HYV corn and HYV rice fertilized at rates 0, 4, 8, and 12 g N m^{-2} in the first cycle of the experiment in 2010. Fertilizer was applied in soil prior to the sowing of seeds of bush bean (cv-MKB 1), long bean (cv-MKP 5), mung bean (cv. local), winged bean (cv. local), and corn. Corn was used as non-N$_2$-fixing reference plant for estimation of N$_2$ fixation by N difference method (NDF). In addition, 16 fallow pots were assigned to fulfill the requirement of rice after fallow crop rotation. Each crop was tested in an individual experiment and each experiment was conducted under completely randomized design with four replications, which covered 16 pots for each crop and a total of 96 pots for all the crop cycles. Rice was transplanted as the 2nd crop after harvesting of first crop cycles of corn and or incorporation of legume residues. After completion of the 1st cycle of rice crop, all legumes and corn were grown in the same pot as third crop in the 2nd cycle. No N fertilizer or other chemical fertilizers were applied in the 2nd year to estimate the enduring effect of legume residues for the next crop. Concurrently, fallow pots were used as rice after fallow crop rotation cycle. Bush bean, long bean, mung bean, winged bean, and corn were planted in early March of 2010 and 2011. All legume crops and corn were harvested at 70 days after emergence. Rice seedlings (14 d old) were transplanted during the 2nd week of July for both years. In the 1st cycle, the rice crop was fertilized in three stages: one third before transplanting, one third at tillering, and one third at panicle primordial initiation stages, respectively. Rice was harvested at the stage of physiological maturity during the second week of November in both years.

Above ground plant parts of all legumes were harvested and fragmented into small pieces and spread into the pots and incorporated to a depth of about 8–10 cm into soil with mulching. This was followed by watering and the pot was left stagnant for 30 days to prepare for rice transplanting. Water was applied every alternate day to keep soil moist until physiological maturity of rice plant.

Determination of Total Dry Matter and Nitrogen

After harvesting of each crop randomly, 200 g fresh plant samples were taken and dried to constant weight at 70°C. Dry weight was measured by digital sensitive balance and converted into above ground biomass yield, rice grain yield, and N content for each crop per unit area. Biomass and grain yield were determined from each pot. Total N concentration was determined by micro-Kjeldahl digestion method [34, 35].

Estimation of Biological Nitrogen Fixation and Nitrogen use Efficiency (NUE)

It is well documented that soil and fertilizer are the sources of N for nonfixing crops while sources of N for fixing crops (F) are from the soil, fertilizer, and atmosphere. For nonfixing and fixing crops, the proportions of N from all the available sources can be denoted as follows [36]:

$$\%\text{Ndff}_{NF} + \%\text{Ndfs}_{NF} = 100\%,$$

$$\%\text{Ndff}_{F} + \%\text{Ndfs}_{F} + \%\text{Ndfa}_{F} = 100\%, \tag{1}$$

$$\%\text{Ndfa} = 100 - (\%\text{Ndff}_{F} + \%\text{Ndfs}_{F}),$$

where Ndff_{NF} denotes N derived from fertilizer for nonfixing crops, Ndfs_{F} denotes N derived from soil for nonfixing crops, Ndff_{F} denotes N derived from fertilizer for fixing crops, Ndfs_{F} denotes N derived from soil for fixing crops, and Ndfa_{F} denotes N derived from atmosphere for fixing crops.

The N difference method (NDF) was used to estimate the contributions of BNF to total N accumulation in the legumes [37]. Consider

$$\%\text{Ndfa} = 100\frac{\left[(\text{Legume N} - \text{Reference N})\right]}{(\text{Legume N})}. \tag{2}$$

The following N-efficiency parameters were calculated for each treatment [38–40].

N agronomic efficiency (NAE g g^{-1}) = (grain yield at Nx − grain yield at N0)/applied N at Nx, and N fertilizer recovery efficiency (NRE %) = (N uptake at Nx − N uptake at N0)/applied at Nx.

Data were analyzed following analysis of variance [41] and treatment means were compared based on the least significant difference (LSD) test at the 0.05 probability level.

RESULTS AND DISCUSSION

Legume crop biomass yield and N uptake were influenced significantly with N fertilizer application. Winged bean grown with N fertilizer at rates of 4, 8, and $12\,g\,N\,m^{-2}$ produced greater biomass ($>185\,g\,m^{-2}$) and N uptake ($>6.5\,g\,m^{-2}$) for both years. In 2010, the biomass yield of bush bean was $148–170\,g\,m^{-2}$ (Table 1) with a parallel N accumulation of $5.0–5.9\,g\,m^{-2}$. As shown in Table 2 in 2011, the biomass of bush bean was $140–163\,g\,m^{-2}$ with the comparable N accumulation of $4.7–5.6\,g\,m^{-2}$ (Table 2). Biomass yield and N uptake of bush bean, long bean, and mung bean were slightly lower compared to winged bean. Biomass and N uptake was higher in winged bean compared to the other tested legume crops in both years. All the tested legume crops, other than winged bean, recorded consistently identical biomass yield and N uptake. Nitrogen uptake in the different legume species vary according to the influence of biomass production and N content in the plant tissues. Earlier studies have reported that faba bean produced $>10\,kg$ biomass m^{-2} which recorded $>35\,g\,N\,m^{-2}$ [33]. Other studies have also reported that N uptake by cover crops produced 4.5 to $22.5\,g$ of N per m^{-2} [42–44]. Furthermore, total N accumulation produced by Vicia faba L. plant residues were lower (11.6 to $19.9\,g\,m^{-2}$) than those gained by Vicia villosa Roth (16.4 to $26.4\,g\,m^{-2}$) [23, 44]. In this study, N uptake by winged bean plant residues was apparently higher than other crops. This was probably due to the biomass yield of long bean, bush bean, and mung bean being lower than that of winged bean.

Table 1:.Biomass yield of bush bean, long bean, mung bean, winged bean, and corn as affected by nitrogen fertilizer

Nitrogen (g/m2)		Biomass yield (g m−2)									
		Bush bean		Long bean		Mung bean		Winged bean		Corn*	
2010	2011	2010	2011	2010	2011	2010	2011	2010	2011	2010	2011
0 0*	0	148.2c	140.1c	138.4c	136.8b	134.9c	131.6c	179.8b	169.4b	462.5d	449.5d
2 4*	0	155.7b	149.8b	145.9b	140.7b	145.0b	141.7b	187.3ab	184.7a	537.5c	514.7c
4 8*	0	165.5a	156.4a	155.7a	149.2a	151.5b	148.2b	190.6a	187.6a	579.8b	560.3b
6 12*	0	170.0a	162.9a	161.9a	155.7a	158.3a	157.7a	197.2a	193.2a	635.2a	602.6a

Means followed by the same letters are not significantly different for each treatment mean (P<0.05).

*Denotes N fertilizer rate for corn in 2010.

Table 2: Nitrogen uptake of bush bean, long bean, mung bean, winged bean, and corn as affected by nitrogen fertilizer

Nitrogen (g/m2)		Nitrogen uptake (gm−2)									
		Bush bean		Long bean		Mung bean		Winged bean		Corn*	
2010	2011	2010	2011	2010	2011	2010	2011	2010	2011	2010	2011
0 0*	0	5.0c	4.7c	4.7d	4.6c	4.6c	4.5d	6.2b	5.6b	3.6d	3.4d
2 4*	0	5.3b	5.1b	5.0c	4.8c	5.0b	4.8c	6.5ab	6.3a	4.3c	4.0c
4 8*	0	5.7a	5.3b	5.4b	5.1b	5.2b	5.1b	6.7a	6.6a	4.8b	4.5b
6 12*	0	5.9a	5.6a	5.7a	5.4a	5.5a	5.3a	6.8a	6.9a	5.3a	4.9a

Means followed by the same letters are not significantly different for each treatment mean (P<0.05).

*Denotes N fertilizer rate for corn in 2010.

Nitrogen Fixation and Nitrogen Recovery Efficiency of Legume Crops

Nitrogen fixation and NRE of legume crops were significantly affected by the use of N fertilizer. Nitrogen fixation in winged bean was 22–42% of the total plant N, 10–29% in bush bean, 6–25% in long bean, and 3–24% in mung bean in 2010. Irrespective of N fertilizer used in 2010, N_2 fixation in winged bean was 30–41%, 13–29% in bush bean, 10–25% in long bean, and 9–24% in mung bean in 2011 as determined by the total NDF method (Table 3). The highest N_2 fixation was achieved by winged bean (41-42%) followed by bush bean (29%), long bean (24-25%), and mung bean (22–24%) when all the legume crops were grown without N fertilizer for both years. Earlier studies have shown that N_2 fixation in Cajanus cajan (L.) Millsp. was 44 to 95% [43] while N_2 fixation by Vicia fabaand Vicia villosa was 41% and 78% of total plant N, respectively [44, 45].

Table 3: Nitrogen fixation of bush bean, long bean, mung bean, and winged bean as affected by nitrogen fertilizer

Nitrogen		Nitrogen fixation (%)							
(g/m2)		Bush bean		Long bean		Mung bean		Winged bean	
2010	2011	2010	2011	2010	2011	2010	2011	2010	2011
0	0	29.4a	28.8a	24.4a	25.1a	22.1a	24.1a	42.1a	41.4a
2	0	19.2b	22.2b	14.2b	16.8b	13.3b	18.0b	33.4b	37.4a
4	0	16.5c	16.1c	11.9c	12.5c	9.2c	11.8c	29.4b	31.8b
6	0	9.9d	13.2d	5.8d	9.5d	3.1d	8.6d	21.7c	29.5b

Means followed by the same letters are not significantly different for each treatment mean (P<0.05).

Winged bean grown with 2 or 4 g N m^{-2} obtained appreciably higher NRE in 2010, while NRE was remarkably higher (29%) when winged bean was grown without N fertilizer in 2011 (Table 4). The superior N_2 fixation and NRE of all tested legume crops were directly linked to the lower rate of N fertilizer application. Nitrogen recovery efficiency in long bean was lower in 2011. The lowest NRE was recorded by legume crops treated with the highest rate of N fertilizer used to the preceding rice crop. Our findings have shown that legume residues mixed into rice crop rotation enrich not only to raise yield but also to nurture and ameliorate soil fertility by virtue of their ability to add ample quantities of atmospheric N. Legumes can make a significant contribution to advance soil fertility and improve soil texture [46–48]. Biomass yield, legume N demand, the capacity to fix N_2, and adaptability to specific environments are important attributes for BNF [49]. Nitrogen content in legume above ground biomass ranged from $4.5\,g\,N\,m^{-2}$ to $6.9\,g\,N\,m^{-2}$, which was integrated into the soil. Among the tested legume plants, winged bean supplied substantially higher N over their residues in each season. The higher quantities of nitrogen N_2 secured in winged bean came from its larger supply of biomass production as well as a higher number of nodules.

Table 4: Nitrogen recovery efficiency of bush bean, long bean, mung bean, and winged bean as affected by nitrogen fertilizer

Nitrogen	Nitrogen recovery efficiency (%)							
(g/m2)	Bush bean		Long bean		Mung bean		Winged bean	
2010	2010	2011	2010	2011	2010	2011	2010	2011
0	0.0c	0.0c	0.0c	0.0c	0.0c	0.0d	0.0d	0.0d
2	14.0b	18.0a	15.0b	13.0b	19.5a	19.5a	15.5a	29.0a
4	16.3a	15.3b	17.0a	15.5a	16.5b	16.0b	14.5b	20.5b
6	14.8b	14.8b	15.7b	14.8a	15.7b	15.0c	11.2c	19.5c

Means followed by the same letters are not significantly different for each treatment mean (P<0.05).

Biomass Yield of Rice after Legume Crops

Biomass yield of rice was strongly influenced by the treatment variables. Significantly greater biomass production was produced by rice after winged bean with 4, 8, and $12\,g\,N\,m^{-2}$ in both years (Table 5). Biomass yield was lower when rice was rotated with other legume crops than in rice after winged bean rotation, but it was higher than in rice after corn rotation. Rice after fallow with 8 and $12\,g\,N\,m^{-2}$ also produced appreciably greater biomass and it was almost at par with rice after bush bean, rice after long bean, or mung bean although the level was comparably lower than rice after winged bean. The lowest biomass yield was produced by rice after corn or rice after fallow grown without N fertilizer (Table 5). The accumulated biomass following rotation with winged bean is possibly an indication of N contribution from the above and below ground residues of the legume. On the contrary, a considerable amount of urea was used and apparently volatile loss of ammonia occurred in rice after fallow rotation with fertilizer. A possible reason this was due to the N fertilizer applied up to panicle initiation stage. Earlier studies have suggested that an appreciable amount of N can be lost via ammonia volatilization when urea was applied at sampling stage just before top dressing [50]. The presence of soil organic matter and decomposed plant residues are the primary determinants of total plant-available N supply for plant growth which is controlled by the balance between N immobilization and mineralization as mediated by soil biota as well as the contributions from applied organic and inorganic N sources and losses from the plant-available N pool [51].

Table 5: Biomass yield of rice as affected by N fertilizer and legume residue

N (g m−2)		Biomass yield (g m−2)					
		Rice after bush bean		Rice after long bean		Rice after mung bean	
2010	2011	2010	2011	2010	2011	2010	2011
0	0	986.0c	921.8c	948.5c	928.7c	952.8c	912.1c
4	0	1058.6b	993.5b	1069.4b	1037.5b	1066.1b	986.9b
8	0	1074.9ab	1074.9a	1135.5ab	1083.4a	1128.7a	1071.7a
12	0	1123.8a	1084.0a	1138.4a	1102.6a	1131.9a	1078.2a
N (g m−2)		Biomass yield (g m−2)					
		Rice after winged bean		Rice after corn		Rice after fallow	
2010	2011	2010	2011	2010	2011	2010	2011
0	0	1154.7b	1107.5b	905.5d	873.0d	941.4c	905.5c
4	0	1262.2a	1211.7a	931.6c	905.5c	1065.1b	993.5b
8	0	1283.4a	1244.3a	1026.1b	960.9b	1136.8a	1042.3a
12	0	1289.9a	1254.1a	1058.6a	1006.5a	1127.0a	1058.6a

Means followed by the same letters are not significantly different for each treatment mean (P<0.05).

Nitrogen Uptake and Nitrogen Efficiency of Rice after Legume Crops

Total N uptake was influenced by N fertilizer application in each crop rotation (Table 6). Maximum N uptake was observed in rice rotation with winged bean with 4, 8, and 12 g N m^{-2}. Rice after long bean or bush bean or mung bean with 8 and 12 g N m^{-2} obtained appreciably greater N uptake compared to rice after fallow or corn. Rice after fallow or in rotation with corn grown without N fertilizer application recorded the least N uptake (Table 6). The increased plant N following rotation with winged bean could probably be attributed to the N contribution from the addition of larger amounts of above ground plant parts and below ground plant residues of legume. The quality of residues in legume plants in rotation with rice crops might be the reason for the greater N uptake compared to rice after corn and rice after fallow rotation. Despite the smaller below ground residues of other pulses other than cereals and other crops, they possess a higher microbial population in the soil which influences the concentrations of nutrients released in the rhizosphere for N uptake by the plants [52].

Table 6: Nitrogen uptake of rice as affected by N fertilizer and legume residue

N (g m^{-2})		Nitrogen uptake (g m^{-2})					
		Rice after bush bean		Rice after long bean		Rice after mung bean	
2010	2011	2010	2011	2010	2011	2010	2011
0	0	9.4d	8.5c	9.3c	8.5c	9.1c	8.4c
4	0	10.4c	9.5b	10.4b	9.5b	10.2b	9.4b
8	0	11.0b	10.7a	11.5a	10.3a	11.4a	10.7a
12	0	11.6a	10.9a	11.6a	10.8a	11.5a	10.8a
N (g m^{-2})		Nitrogen uptake (g m^{-2})					
		Rice after winged bean		Rice after corn		Rice after fallow	
2010	2011	2010	2011	2010	2011	2010	2011
0	0	12.1b	11.3b	7.1d	6.8d	7.5c	7.1d
4	0	13.4a	12.5a	7.6c	7.2c	8.5b	7.9c
8	0	13.7a	12.8a	8.4b	7.7b	9.5a	8.4b
12	0	13.9a	12.9a	8.9a	8.3a	9.7a	8.7a

Means followed by the same letters are not significantly different for each treatment mean (P<0.05).

Among crop rotations, rice after winged bean with N at rates of 4, 8, or 12 g m^{-2} showed the highest N uptake, whilst rice after other legumes with N at rates of 4, 8, or 12 g m^{-2} recorded intermediate N uptake and rice after corn and rice after fallow rotation recorded the lowest. Soil microbial population increased rapidly when young and relatively succulent green manure crop are incorporated into the soil. The soil microorganisms multiply faster to invade the freshly incorporated plant residues. After decomposition of plant residues through microbial breakdown, nutrients remain within the plant tissues and nutrient rich dead microbes, which are released and made available to the following crop [51]. In our study, all legume crops were added into soil at the pod formation stage. Apparently, this could slightly slow the breakdown of legume residues resulting in poor volatile loss of ammonia during rice after winged bean or rice in rotation with other legume growing cycles. While in rice fallow systems with fertilizer, a substantial amount of the applied urea are seemingly lost by ammonia volatilization.

Nitrogen recovery efficiency (NRE) was influenced significantly by N fertilizer application in each crop rotation (Table 7). The highest NRE (32.5%) was obtained by rice after winged bean with 4 g N m^{-2}. In this study, the observed NRE values were relatively lower than those reported in previous studies where N fertilizer was used later in the crop growing period. The NRE was significantly higher in 2010 than in 2011, possibly due to the fact that the highest grain yield was recorded in 2010 (Table 7). The higher grain yield of rice in 2010 could be due to higher N uptake, caused by both legume residual effects along with N fertilizer. Regardless of fertilizer, rice after legumes obtained higher NRE than rice after fallow or corn in both years. In India, the large variation in NRE (18%—1st year and 49%—2nd year) was observed in rice-wheat systems. This difference was directly linked with poor yields in the first year caused by unfavorable weather and highlights the importance of higher crop growth and yield to higher NRE [49]. Higher NRE values were reported in rotation than in monoculture [40, 53]. Regardless of N fertilizer rate, the NRE values were 15–33% in 2010 and 13–30% in 2011 for rice after winged bean (Table 7). The NRE values were 20–29% in 2010 and 20–28% in 2011 for rice after bush bean or long bean or mung bean. In our study, the NRE values were lower compared to NRE values (42%) obtained in developed countries [50]. A possible reason could be due to the fact that the present study was carried out under greenhouse conditions which does not show the full potential of legume performance while in developed countries studies were conducted in on-station field conditions.

Table 7: Nitrogen recovery efficiency of rice as affected by N fertilizer and legume residue

N (g m−2)		Nitrogen recovery efficiency (%)					
		Rice after bush bean		Rice after long bean		Rice after mung bean	
2010	2011	2010	2011	2010	2011	2010	2011
0	0	0.0c	0.0d	0.0c	0.0d	0.0c	0.0d
4	0	28.8a	27.5a	28.8a	27.5a	27.5a	25.0b
8	0	29.5a	23.8b	29.4a	23.8b	28.8a	28.8a
12	0	19.6b	20.0c	19.6b	20.0c	20.0b	20.0c
N (g m−2)		Nitrogen recovery efficiency (%)					
		Rice after winged bean		Rice after corn		Rice after fallow	
2010	2011	2010	2011	2010	2011	2010	2011
0	0	0.0d	0.0d	0.0c	0.0c	0.0c	0.0d
4	0	32.5a	30.0a	10.0b	9.8b	25.5a	20.0a
8	0	20.0b	18.8b	16.3a	11.1a	25.0a	16.3b
12	0	15.0c	13.3c	15.0a	12.4a	18.3b	13.3c

Means followed by the same letters are not significantly different for each treatment mean (P<0.05).

The higher N recoveries in legume rotations could be due to the enrichment of soil N. The below ground pool of legume N is an important source of N for subsequent crops [51]. When the soil N content rises, the amount of sequestered N contributes to a greater NUE of the cropping system and the amount of sequestered N achieved from applied N results in a higher NRE [51]. The average NUE gained by rice farmers is 31% of applied N based upon on-farm estimation in the major rice production countries of Asia. In contrast, it is documented that the recovery efficiency of nitrogen for rice normally varies between 50 and 80% in well-managed field experiments [51]. Therefore, emphasis has been given to improve recovery efficiency of nitrogen because the N fertilizer is the greatest source of N input and loss from cereal cropping systems.

Nitrogen agronomic efficiency (NAE) was affected significantly by N fertilizer application in each crop rotation. Rice after winged bean grown with $4\,g\,N\,m^{-2}$ recorded the highest NAE ($24\,g\,g^{-1}$ to $27\,g\,g^{-1}$) for both years (Table 8). Rice after long bean with $4\,g\,N\,m^{-2}$also showed a similar trend although NAE was lower than rice after winged bean systems regardless of N fertilizer for both years (Table 8). The NAE trends were similar in both 2010 and 2011. Rice after fallow and rice after long bean with $4\,g\,N\,m^{-2}$ also showed similar trends although NAE credentials of rice after winged bean systems indicate a positive response to rice production without deteriorating soil fertility. Therefore, the nitrogen agronomic efficiency and grain yield was significantly increased by the legume plant residues when supported by organic sources of N. The results showed that the increase in the application of N caused the decline of agronomic nitrogen efficiency. The highest and the lowest agronomical nitrogen use efficiencies were obtained in the 60 and $180\,kg\,N\,ha^{-1}$ which showed that NAE decreased with the increasing rate of N fertilizer used [54]. From this, it can be said that the fertilizer response to NRE was poor but grain yield and N uptake showed significant differences among fertilizer rates. In this regard, NUE was reduced at higher rates of fertilizer N rate, possibly due to greater losses from soil via volatilization or leaching losses [53]. With respect to fertilizer utilization in rice crop, current N fertilizer management strategies must be improved in upland as well as wetland rice growing areas in Malaysia.

Table 8: Nitrogen agronomic efficiency ($g\,g^{-1}$) of rice as affected by N fertilizer and legume residues

N (g m−2)		Nitrogen agronomic efficiency (g g−1)					
		Rice after bush bean		Rice after long bean		Rice after mung bean	
2010	2011	2010	2011	2010	2011	2010	2011
0	0	0.0d	0.0d	0.0d	0.0c	0.0d	0.0d
4	0	13.0a	12.5a	21.2a	17.1a	14.8a	12.9a
8	0	10.7b	9.7b	15.1b	11.4b	12.6b	11.4b
12	0	7.4c	6.5c	12.8c	11.1b	8.3c	7.8c
N (g m−2)		Nitrogen agronomic efficiency (g g−1)					

		Rice after winged bean		Rice after corn		Rice after fallow	
2010	2011	2010	2011	2010	2011	2010	2011
0	0	0.0d	0.0d	0.0c	0.0c	0.0	0.0d
4	0	26.9a	23.6a	14.7a	16.3a	21.2a	13.8b
8	0	16.3b	14.7b	15.5a	14.3b	18.7b	15.1a
12	0	12.2c	12.5c	13.3b	12.2b	15.2c	11.4c

Means followed by the same letters are not significantly different for each treatment mean (P<0.05).

Grain Yield and Harvest Indices of Rice

Rice grain yield increased significantly by the amendment of soil with addition of legume residues and N fertilizer application (Table 9). Maximum grain yield (603–684 gm^{-2}) was produced by rice after winged bean with 4, 8, and 12 g N m^{-2} in both years. Rice in rotation with corn (293–349 g m^{-2}) and rice after fallow (342–371 g m^{-2}) grown without fertilizer recorded the lowest yield of rice crops. Rice after long bean or bush bean or mung bean with 8 and 12 g N m^{-2} gave similar yields for both years. In 2011, a slightly lower yield was obtained but a similar trend was observed in all the crop rotation systems. Rice in rotation with corn grown with 8 g N m^{-2} and 12 g N m^{-2} produced comparatively lower yield than other counterparts. Identical yield was observed in rice after fallow with 8 g N m^{-2} and 12 g N m^{-2} (Table 9). In rice after fallow rotation, rice yield showed a positive response to fertilizer rates even at 12 g N m^{-2}, which suggested that higher yields could be produced with higher rates of N fertilizer application. In both years, legume crop residues incorporation in rice after winged bean and rice after long bean crop rotation systems was effective in producing a satisfactory yield even in 2011 when rice after winged bean was grown without N fertilizer. It was observed that rice grain yield decreased about 5–33% in the zero-N control among legume crop rotation in the second year of experiment. The rice grain yield proved that winged bean was more effective than N fertilizer for both years. Similar results were obtained in rice after hairy vetch with 4 or 8 g N m^{-2} [44]. In 2011, rice grain yield was slightly lower when grown without N fertilizer in the case of rice after winged bean rotation, but the addition of winged bean residues contributed noticeably to higher yield levels when compared with rice after fallow with 8 or 12 g N m⁻². The present findings differed with observations from other studies [18], which have reported that rice yield did not increase or even decrease when rice straw residue was added without fertilizer N. The possible reason for this could be due to the use of rice straw residues under upland conditions which were not fully decomposed during the rice growing season. Additionally, grain yield of rice significantly increased when residues were incorporated into flooded soil [47]. Our results suggest that legume residues had a positive influence on both rice yield and N uptake when fertilizer was not used. Certainly, the influence of legume residues on rice yield depends on the soil nutrient status, texture, addition of organic matter, amount of residue returned to soil [33], and timing and levels of fertilizer N used.

Table 9: Grain yield of rice as affected by N fertilizer and legume residues

N (g m−2)		Yield (g m−2)					
		Rice after bush bean		Rice after long bean		Rice after mung bean	
2010	2011	2010	2011	2010	2011	2010	2011
0	0	424.5c	358.3c	416.9c	403.9c	407.2c	395.7c
		(−21)	(−33)	(−22)	(−25)	(−24)	(−26)
4	0	495.1b	439.7b	501.6b	472.3b	495.1b	449.2b
		(−7)	(−18)	(−6)	(−12)	(−7)	(−16)
8	0	547.2a	472.3ab	537.5a	495.1ab	540.0a	488.7a
		(+2)	(−12)	(+1)	(−7)	(+1)	(−8)
12	0	553.7a	521.2a	570.0a	537.5a	565.3a	521.2a
		(+4)	(−2)	(+7)	(+1)	(+6)	(−2)
N (g m−2)		Yield (g m−2)					
		Rice after winged bean		Rice after corn		Rice after fallow	
2010	2011	2010	2011	2010	2011	2010	2011
0	0	537.5b	508.1b	348.5d	293.2d	371.3c	342.0c
		(+1)	(−5)	(−35)	(−45)	(−31)	(−36)
4	0	645.0a	602.6a	407.2c	358.3c	488.6b	423.5b
		(+21)	(+13)	(−24)	(−33)	(−8)	(−21)
8	0	667.8a	625.4a	472.3b	407.2b		472.3a
		(+25)	(+17)	(−12)	(−24)	(100)	(−12)
12	0	684.0a	658.0a	508.1a	439.7a	553.7a	504.9a

		(+28)	(+23)	(−5)	(−18)	(+4)	(−5)

Means followed by the same letters are not significantly different for each treatment means (P<0.05).

*Parenthesis values denote yield increase (+) or decrease (−) in percent; values are calculated based on rice after fallow with $8\,g\,N\,m^{-2}$ (100%).

Addition of legume residues and N fertilizer application had a significant effect on harvest indices (HI) for both years (Table 10). In both 2010 and 2011, rice crop rotation with winged bean grown with 4, 8, and $12\,g\,N\,m^{-2}$, achieved higher HI compared to the other crop rotation systems. Rice after fallow or corn gown without fertilizer N recorded the lowest HI for both years. Rice after bush bean with 8 or $12\,g\,N\,m^{-2}$ gave superior HI in 2010 and similarly rice after long bean or mung bean with $12\,g\,N\,m^{-2}$ gave better HI in 2010. Rice after fallow or corn with $12\,g\,N\,m^{-2}$ showed identical HI for both years (Table10). The harvest indices were lower in all systems in 2011 except for rice after winged bean. It has been documented that harvest indices increases significantly with higher rates of N fertilizer application [55]. However, on the contrary, other studies have reported that with increasing N fertilizer, the harvest index decreased [56]. The lower HI with higher N fertilizer application suggested that N fertilizer influenced more biological yield than economic yield [57]. The grain harvest index values were 43% at zero N level and 50% at $400\,mg\,N\,kg^{-1}$ across 19 upland rice genotypes [16]. Nitrogen significantly improved grain harvest index, nitrogen harvest index and plant height which are positively associated with grain yield [58, 59]. The higher levels of phosphorus ($72\,kg\,ha^{-1}$) application in rice also recorded higher harvest index [60]. Harvest index is a measure of success in partitioning assimilated photosynthate. An improvement of harvest index means an increase in the economic portion of the plant [61]. Our observations suggest that higher fertilizer N application and their residual effect along with N_2 fixation from legume crops influence HI in legume crop rotation systems.

Table 10: Harvest indices of rice as affected by N fertilizer and legume residues

N (gm−2)		Harvest indices					
		Rice after bush bean		Rice after long bean		Rice after mung bean	
2010	2011	2010	2011	2010	2011	2010	2011
0	0	42.9c	38.9c	43.5c	43.5c	42.7c	43.2b
4	0	46.8b	44.3b	45.6b	45.5b	46.4b	45.5b
8	0	50.9a	43.9b	45.7b	45.7b	47.6b	45.6b
12	0	49.3a	48.1a	48.7a	48.7a	49.5a	48.3a
N (gm−2)		Harvest indices (%)					
		Rice after winged bean		Rice after corn		Rice after fallow	
2010	2011	2010	2011	2010	2011	2010	2011
0	0	46.6b	45.9b	38.5c	33.2c	39.4d	35.3c
4	0	51.1a	49.7a	43.7b	39.6b	41.4c	37.7b
8	0	52.0a	50.3a	46.0a	42.4a	45.8b	42.2a
12	0	53.0a	52.5a	48.0a	43.7a	49.1a	

Means followed by the same letters are not significantly different for each treatment mean (P<0.05).

CONCLUSIONS

The present study has shown that both legume residues and N fertilizer affected grain yield, N uptake, NAE, and NRE, which were greater in 2010 than in 2011. Rice rotation by all legumes produced consistently higher grain yield and NUE than rice after fallow. However, the NUE declined with higher levels of fertilizer N used, reflecting poor N utilization by the rice crop. This indicated that though rice rotation with legume crops plays a significant role in the improvement of grain yield, higher levels can be sustained by compatible and proper management of residues and N fertilizer. The N difference method applied in this study exhibited that N produced by all the legumes was readily available and can be used efficiently by the rice crop. Winged bean was capable of producing greater amount of biomass and providing high quantities of total N, in addition to fixing substantial quantities of

N. Without significant loss of yield level, winged bean plant residue incorporation can be an alternative source to N fertilizer for sustainable rice yield. However, the incorporation of long bean plant residues requires minimum N fertilizer ($4\,g\,N\,m^{-2}$) and can be an alternative to the sole use of N fertilizer while bush bean and mung bean plant residues along with N fertilizer ($8\,g\,N\,m^{-2}$) can also be an alternative to N management method to reduce N losses from N fertilizer applied to rice crop. Winged bean plant residues are able to provide sufficient N to the soil for the rice crop and afford an advantage equivalent to that of 4 to $8\,g$ fertilizer $N\,m^{-2}$, respectively. In conclusion, amongst the tested legumes, winged bean showed the greatest potential while the other legumes can also be used as a substitute or supplement in place of chemical/inorganic N fertilizers.

ACKNOWLEDGMENTS

The authors thank the University of Malaya for their generous financial support (RG088/10SUS). The technical assistance of the Department of Chemistry personnel's, Laboratory for Chemical Analysis and to avail their laboratory facilities is highly appreciated.

REFERENCES

1.   A. Makino, "Photosynthesis, grain yield, and nitrogen utilization in rice and wheat,"Plant Physiology, vol. 155, no. 1, pp. 125–129, 2011.

2.   T. Mayumi, A. Tomomi, and Y. Tomoyuki, "Assimilation of ammonium ions and reutilization of nitrogen in rice (Oryza sativa L.)," Journal of Experimental Botany, vol. 58, no. 9, pp. 2319–2327, 2007.

3.   S. M. Haefele, S. M. A. Jabbar, J. D. L. C. Siopongco et al., "Nitrogen use efficiency in selected rice (Oryza sativa L.) genotypes under different water regimes and nitrogen levels," Field Crops Research, vol. 107, no. 2, pp. 137–146, 2008.

4.   S. M. Haefele, K. Naklang, D. Harnpichitvitaya et al., "Factors affecting rice yield and fertilizer response in rainfed lowlands of northeast Thailand," Field Crops Research, vol. 98, no. 1, pp. 39–51, 2006.

5. N. K. Ho, S. Jegatheesan, and F. K. Phang, "Increasing rice productivity in Malaysia-an independent view," in Proceedings of the National Conference & Workshop on Food Security, Kuala Lumpur, Malaysia, December 2008.

6. X. R. Wei, M. D. Hao, M. Shao, and W. J. Gale, "Changes in soil properties and the availability of soil micronutrients after 18 years of cropping and fertilization," Soil and Tillage Research, vol. 91, pp. 120–130, 2006.

7. J. K. Ladha and P. M. Reddy, "Nitrogen fixation in rice systems: state of knowledge and future prospects," Plant and Soil, vol. 252, no. 1, pp. 151–167, 2003.

8. S. L. Zhang, Y. A. Tong, D. L. Liang, D. Q. Lu, and E. Ove, "Nitrate N movement in the soil profile as influenced by rate and timing of nitrogen application," Acta Pedologica Sinica, vol. 41, pp. 270–277, 2004 (Chinese).

9. J. Richter and M. Roelcke, "The N-cycle as determined by intensive agriculture—examples from central Europe and China," Nutrient Cycling in Agroecosystems, vol. 57, no. 1, pp. 33–46, 2000.

10. G. X. Xing and Z. L. Zhu, "An assessment of N loss from agricultural fields to the environment in China," Nutrient Cycling in Agroecosystems, vol. 57, no. 1, pp. 67–73, 2000.

11. J. G. Zhu, Y. Han, G. Liu, Y. L. Zhang, and X. H. Shao, "Nitrogen in percolation water in paddy fields with a rice/wheat rotation," Nutrient Cycling in Agroecosystems, vol. 57, no. 1, pp. 75–82, 2000.

12. C. Kirda, M. R. Derici, and J. S. Schepers, "Yield response and N-fertiliser recovery of rainfed wheat growing in the Mediterranean region," Field Crops Research, vol. 71, no. 2, pp. 113–122, 2001.

13. G. Wang, A. Dobermann, C. Witt, Q. Sun, and R. Fu, "Performance of site-specific nutrient management for irrigated rice in Southeast China," Agronomy Journal, vol. 93, no. 4, pp. 869–878, 2001.

14. W. B. Stevens, R. G. Hoeft, and R. L. Mulvaney, "Fate of nitrogen[-15] in a long-term nitrogen rate study: II. Nitrogen uptake efficiency," Agronomy Journal, vol. 97, no. 4, pp. 1046–1053, 2005.

15. S. Delin, A. Nyberg, B. Lindén et al., "Impact of crop protection on nitrogen utilisation and losses in winter wheat production," European Journal of Agronomy, vol. 28, no. 3, pp. 361–370, 2008.

16. N. K. Fageria and V. C. Baligar, "Enhancing nitrogen use efficiency in crop plants,"Advances in Agronomy, vol. 88, pp. 97–185, 2005.

17. N. K. Fageria, N. A. Slaton, and V. C. Baligar, "Nutrient management for improving lowland rice productivity and sustainability," Advances in Agronomy, vol. 80, pp. 64–152, 2003.

18. N. Thuy, H. Shan, S. Bijay, et al., "Nitrogen supply in rice-based cropping systems as affected by crop residue management," Soil Science Society of America Journal, vol. 72, no. 2, pp. 514–523, 2008.

19. E. Cazzato, V. Laudadio, A. M. Stellacci, E. Ceci, and V. Tufarelli, "Influence of sulphur application on protein quality, fatty acid composition and nitrogen fixation of white lupin (Lupinus albus L)," European Food Research and Technology, vol. 235, no. 5, pp. 963–969, 2012.

20. D. K. Kundu and J. K. Ladha, "Efficient management of soil and biologically fixed N2 in intensively-cultivated rice fields," Soil Biology and Biochemistry, vol. 27, no. 4-5, pp. 431–439, 1995.

21. S. M. Dabney, J. A. Delgado, and D. W. Reeves, "Using winter cover crops to improve soil and water quality," Communications in Soil Science and Plant Analysis, vol. 32, no. 7-8, pp. 1221–1250, 2001.

22. S. Yadvinder, S. Bijay, and J. Timsina, "Crop residue management for nutrient cycling and improving soil productivity in rice-based cropping systems in the tropics," Advances in Agronomy, vol. 85, pp. 269–407, 2005.

23. I. Rochester and M. Peoples, "Growing vetches (Vicia villosa Roth) in irrigated cotton systems: inputs of fixed N, N fertiliser savings and cotton productivity," Plant and Soil, vol. 271, no. 1-2, pp. 251–264, 2005.

24. M. S. Ali, "Evolution of fertilizer use by crops in : recent trends and prospects," inProceedings of the International Fertilizer Industry Association, p. 3, IFA Cross Roads, Asia-Pacific, Kota Kinabalu, Malaysia, December 2009.

25. A. B. N. M. Wan, "Country paper: Malaysia, impact of land utilization systems on agricultural productivity," in Report of the APO Seminar, pp. 226–240, APO Japan 2003, 2000, Islamic Republic of Iran, 4–9 November 2000.

26. A. Faridah, "Sustainable agriculture system in Malaysia," in Proceedings of the Regional Workshop on Integrated Plant Nutrition System (IPNS '01), Development in Rural Poverty Alleviation, UN Conference Complex, Bangkok, Thailand, September 2001.

27. A. R. Khairuddin, "Biofertilizers in Malaysian agriculture: perception, demand and promotion. Country Report of Malaysia," in Proceedings of the FNCA Joint Workshop on Mutation Breeding and Biofertilizer, Beijing, China, August 2002.

28. D. A. Derksen, K. E. McGillivary, S. J. Neudorf, et al., Wheat-Pea Management Study. Annual Report, Agriculture and Agri-Food Canada, Brandon Research Centre, Brandon, Canada, 2001.

29. A. E. Russell, D. A. Laird, and A. P. Mallarino, "Nitrogen fertilization and cropping system impacts on soil quality in midwestern mollisols," Soil Science Society of America Journal, vol. 70, no. 1, pp. 249–255, 2006.

30. K. Kumar and K. M. Goh, "Crop residues and management practices: effects on soil quality, soil nitrogen dynamics, crop yield, and nitrogen recovery," Advances in Agronomy, vol. 68, pp. 197–319, 2000.

31. K. M. Goh, D. R. Pearson, and M. J. Daly, "Soil physical, chemical and biological indicators of soil quality in conventional, biological and integrated apple orchard management systems," Biological Agriculture and Horticulture, vol. 18, pp. 269–292, 2001.

32. K. Kumar, K. M. Goh, W. R. Scott, and C. M. Frampton, "Effects of ^{15}N-labelled crop residues and management practices on subsequent winter wheat yields, nitrogen benefits and recovery under field conditions," Journal of Agricultural Science, vol. 136, no. 1, pp. 35–53, 2001.

33. M. R. Motior, T. Amano, H. Inoue, Y. Matsumoto, and T. Shiraiwa, "Nitrogen uptake and recovery from N fertilizer and legume crops in wetland rice measured by ^{15}N and non-isotope techniques," Journal of Plant Nutrition, vol. 34, no. 3, pp. 402–426, 2011.

34. G. X. Chu, Q. R. Shen, and J. L. Cao, "Nitrogen fixation and N transfer from peanut to rice cultivated in aerobic soil in an intercropping system and its effect on soil N fertility," Plant and Soil, vol. 263, no. 1-2, pp. 17–27, 2004.

35. J. M. Bremner, "Nitrogen-total," in Methods of Soil Analysis. Part 3: Chemical Methods, D. L. Sparks, Ed., pp. 1085–1121, Soil Science Society of America Inc, American Society of Agronomy Inc, Madison, Wis, USA, 1996.

36. IAEA, A Manual: Use of Isotope and Radiation Methods in Soil and Water Management and Crop Nutrition, FAO/IAEA Agriculture and Biotechnology Laboratory Agency's Laboratories, Seibersdorf and Soil and water Management and Crop Nutrition Section, International Atomic Energy Agency, Vienna, Austria, 2001.

37. M. B. Peoples, R. M. Boddey, and D. F. Herridge, "Quantification of nitrogen fixation," in Nitrogen Fixation at the Millennium, G. J. Leigh, Ed., pp. 357–389, Elsevier Science, Amsterdam, The Netherlands, 2002.

38. K. G. Cassman, A. Dobermann, and D. T. Walters, "Agroecosystems, nitrogen-use efficiency, and nitrogen management," Ambio, vol. 31, no. 2, pp. 132–140, 2002.

39. R. J. López-Bellido and L. López-Bellido, "Efficiency of nitrogen in wheat under Mediterranean conditions: effect of tillage, crop rotation and N fertilization," Field Crops Research, vol. 71, no. 1, pp. 31–46, 2001.

40. Z. Shi, D. Li, Q. Jing et al., "Effects of nitrogen applications on soil nitrogen balance and nitrogen utilization of winter wheat in a rice-wheat rotation," Field Crops Research, vol. 127, pp. 241–247, 2012.

41. SAS, "SAS Institute Incorporated, Release 9.1," Cary, NC, USA, 2008.

42. J. Evans, A. M. McNeil, M. J. Unkovich, N. A. Fettel, and D. P. Heenam, "Net nitrogen balances for cool-season grain legume crops and contributions to wheat nitrogen uptake: a review," Australian Journal of Experimental Agriculture, vol. 41, no. 3, pp. 347–359, 2001.

43. M. B. Peoples and D. F. Herridge, "Nitrogen fixation by legumes in tropical and subtropical agriculture," Advances in Agronomy, vol. 44, pp. 155–223, 1990.

44. M. M. Rahman, T. Amano, and T. Shiraiwa, "Nitrogen use efficiency and recovery from N fertilizer under rice-based cropping systems," Australian Journal of Crop Science, vol. 3, no. 6, pp. 336–351, 2009.

45. E. Cazzato, V. Tufarelli, E. Ceci, A. M. Stellacci, and V. Laudadio, "Quality, yield and nitrogen fixation of faba bean seeds as affected by sulphur fertilization," Acta Agriculturae Scandinavica B: Soil Plant Science, vol. 62, no. 8, pp. 732–738, 2012.

46. P. Sullivan, "Overview of cover crops and green manures fundamentals of sustainable agriculture," ATTRA-National Sustainable Agriculture Information Service, Fayetteville, Ark, USA, 2003, http://www.attra.ncat.org/attra-pub/PDF/covercrop.pdf.

47. B. Singh, Y. H. Shan, S. E. Johnson-Beebout, Y.-S. Yadvinder-Singh, and R. J. Buresh, "Crop residue management for lowland rice-based cropping systems in Asia," Advances in Agronomy, vol. 98, pp. 117–199, 2008.

48. K. H. Diekmann, S. K. de Datta, and J. C. G. Ottow, "Nitrogen uptake and recovery from urea and green manure in lowland rice measured by ^{15}N and non-isotope techniques," Plant and Soil, vol. 148, no. 1, pp. 91–99, 1993.

49. E. L. Balota, A. Colozzi-Filho, D. S. Andrade, and R. P. Dick, "Microbial biomass in soils under different tillage and crop rotation systems," Biology and Fertility of Soils, vol. 38, no. 1, pp. 15–20, 2003.

50. W. R. Raun and G. V. Johnson, "Improving nitrogen use efficiency for cereal production," Agronomy Journal, vol. 91, no. 3, pp. 357–363, 1999.

51. F. Wichern, E. Eberhardt, J. Mayer, R. G. Joergensen, and T. Müller, "Nitrogen rhizodeposition in agricultural crops: methods, estimates and future prospects," Soil Biology and Biochemistry, vol. 40, no. 1, pp. 30–48, 2008.

52. D. Dawe, A. Dobermann, J. K. Ladha et al., "Do organic amendments improve yield trends and profitability in intensive rice systems?" Field Crops Research, vol. 83, no. 2, pp. 191–213, 2003.

53. D. R. Huggins and W. L. Pan, "Key indicators for assessing nitrogen use efficiency in cereal-based agroecosystems," Journal of Crop Production, vol. 8, no. 1-2, pp. 157–185, 2003.

54. J. S. Shahzad, Z. M. Roghayyeh, Y. Asgar, K. Majid, and G. Roza, "Study of agronomical nitrogen use efficiency of durum wheat

affected by nitrogen fertilizer and plant density," World Applied Sciences, vol. 11, no. 6, pp. 674–681, 2010.

55.   K. M. Panahyan and S. S. H. Jamaati, "Response of phenology and dry matter remobilization of durum wheat to nitrogen and plant density," World Applied Sciences Journal, vol. 10, no. 3, pp. 304–310, 2010.

56.   A. Hamidi and A. M. Dabagh, "Study of grain yield and its components, biomass and harvest index of two hybrid of corn under plant densities and nitrogen fertilizer levels," Iranian Journal of Agricultural Sciences, vol. 10, no. 1, pp. 39–35, 1995.

57.   N. H. Hazeri, A. Tobeh, A. Gholipouri, H. Mostafaei, and S. S. H. Jamaati, "The effect of nitrogen and phosphorous rates on fertilizer use efficiency in lentil," World Applied Sciences Journal, vol. 9, no. 9, pp. 1043–1046, 2010.

58.   N. K. Fageria, "Yield physiology of rice," Journal of Plant Nutrition, vol. 30, no. 6, pp. 843–879, 2007.

59.   N. K. Fageria, A. Moreira, and A. M. Coelho, "Yield and yield components of upland rice as influenced by nitrogen sources," Journal of Plant Nutrition, vol. 34, no. 3, pp. 361–370, 2011.

60.   M. M. Alam, M. H. Ali, A. K. M. Ruhul Amin, and M. Hasanuzzaman, "Yield attributes, yield and harvest index of three irrigated rice varieties under different level of phosphous," Advances in Biological Research, vol. 3, no. 3-4, pp. 132–139, 2009.

61.   X. Li, W. Yan, H. Agrama et al., "Unraveling the complex trait of harvest index with association mapping in rice (Oryza sativa L.)," PLoS ONE, vol. 7, no. 1, Article ID e29350, 2012.

Chapter 7

Tillage System Affects Soil Organic Carbon Storage and Quality in Central Morocco

R. Moussadek[1], R. Mrabet[1], R. Dahan[1], A. Zouahri[1],
M. El Mourid[2], and E. Van Ranst[3]

[1]Institut National de la Recherche Agronomique, BP 415, Avenue la Victoire, Rabat, Morocco

[2]International Center for Agricultural Research in the Dry Areas, Menzah IV, 2037 Tunis, Tunisia

[3]Laboratory of Soil Science, Department of Geology and Soil Science (WE13), Ghent University, Krijgslaan 281/S8, 9000 Gent, Belgium

ABSTRACT

Stabilizing or improving soil organic carbon content is essential for sustainable crop production under changing climate conditions. Therefore, soil organic carbon research is gaining momentum in the

Mediterranean basin. Our objective is to quantify effects of no tillage (NT) and conventional tillage (CT) on soil organic carbon stock (SOCs) in three soil types (Vertisol, Cambisol, and Luvisol) within Central Morocco. Chemical analyses were used to determine how tillage affected various humic substances. Our results showed that, after 5 years, surface horizon (0–30 cm) SOC stocks varied between tillage systems and with soil type. The SOCs was significantly higher in NT compared to CT (10% more in Vertisol and 8% more in Cambisol), but no significant difference was observed in the Luvisol. Average SOCs within the 0–30 cm depth was 29.35 and 27.36 Mg ha^{-1} under NT and CT, respectively. The highest SOCs (31.89 Mg ha^{-1}) was found in Vertisols under NT. A comparison of humic substances showed that humic acids and humin were significantly higher under NT compared to CT, but fulvic acid concentrations were significantly lower. These studies confirm that NT does have beneficial effects on SOCs and quality in these soils.

INTRODUCTION

Management of soil organic matter (SOM) in arable lands has become increasingly important in many areas of the world in order to combat land degradation [1, 2], increase food security [3, 4], reduce C emissions, and/or mitigate climate change [5–7]. In fact, soil carbon cycling and composition are essential components of comprehensive agricultural and ecological impacts and forecasting. Bot and Benites [8] have considered SOM as keys to developing drought-resistant soils (i.e., water conservation, evaporation and erosion control, and soil water infiltration ease) and ensuring sustainable food production (crop productivity, fertilizer use efficiency, reduced pesticide use, and crop ecological intensification) [9, 10].

In Morocco, previous investigations have shown that SOM content in most soils is low (<2%) and the decadal average (from 1987 to 1997) loss of the SOM due to intensive land use is about 30% [6]. The decline of SOM in cultivated soil of Morocco, due to the tillage intensification, decreased soil quality and increased the risk of soil degradation [11, 12]. In fact, FAO estimated that 71% of Moroccan agricultural soils are degraded and require conservation measures [8].

To deal with this situation, conservation agriculture has been recommended as an alternative strategy to invert the soil degradation spiral in many parts of the world [14–16]. No-tillage systems, which consist of eliminating soil tillage and inversion, maintaining crop residue cover, and ensuring proper crop sequences, have been reported to improve SOM level and ensure carbon accumulation and sequestration in diverse soils from contrasted climate regimes [17]. In the Mediterranean basin, experiences have shown the sustainable use of natural resources through adoption and diffusion of no-tillage systems improves soil quality and enhances crop productivity vis-à-vis climate variability and drought [18, 19]. In other terms, SOM improvement under NT is fundamental for food system in Mediterranean drylands. In Spain, Álvaro-Fuentes et al. [20] showed that no tillage (NT) increases soil organic carbon (SOC) stock in the soil profile (0–40 cm) compared to CT. Similar results were found in Italy [21], in France [22], and in Morocco [23–25]. It was also found that, under no-tillage systems, dryland soil can play a part in mitigating CO_2 levels [22, 26]. The continuous harvesting of plant materials in conventional tillage system was reported by several authors of decreasing carbon levels in Mediterranean soils and harming its fertility and health [17]. Although most studies reported that no tillage increases SOC storage in the upper soil horizons compared to conventional tillage, other studies reported that SOC content, in deeper horizons, could be similar or even lower under NT than under conventional tillage system [26]. In fact, Franzluebbers [27] reported that SOC stocks depend not only on tillage system, but also on local conditions (soil type and climate).

SOM is a continuum of substances in all stages of decay. Humus is a relatively stable component formed by humic substances, including humic acids, fulvic acids, hymatomelanic acids, and humins [28]. Humic and fulvic substances enhance plant growth directly through physiological and nutritional effects [8] but also improve soil health and quality through amelioration of soil's physical, biological, and chemical properties. It is evident to account no-tillage systems as durable management practices; they have to improve and protect simultaneously soil organic matter and its active fraction.

Investigations reported that NT systems affect not only the amount of SOM but also its characteristics [29]. Soil organic matter quality is affected by no tillage either in terms of particulate organic matter [24] or in terms of its composition of humic acids, fulvic acids, and humin

[30]. These humic substances are involved in improving soil structural stability and plant growth [31].

In Morocco, the effect of no tillage on SOC content has been largely investigated [32], but studies on the potential storage of SOC under NT, taking into account the effect of soil types, have not been carried out yet. Furthermore, effect of tillage systems on humic substances in semiarid Morocco's environment was not assessed in any previous research. Hence, SOC chemistry remains scarcely studied under tillage systems for the most productive soils in Morocco, as Vertisols. Thus, the objectives of this study are

- to investigate the effect of NT and CT on the SOC stock (SOCs) in three soil types;
- to quantify the effect of NT and CT on the composition of humic substances in Vertisols.

MATERIAL AND METHODS

Research Sites, Experimental Set-Up, and Crop Management

This study was conducted at the Merchouch plateau ($33°34'N$, $6°42'W$, and 425 m elevation), in Morocco. Its Mediterranean climate is characterised by a mean annual rainfall of 450 mm. Figure 1 shows the monthly time-series (1970–2012) of the average rainfall and temperature and those occurring during the studies period.

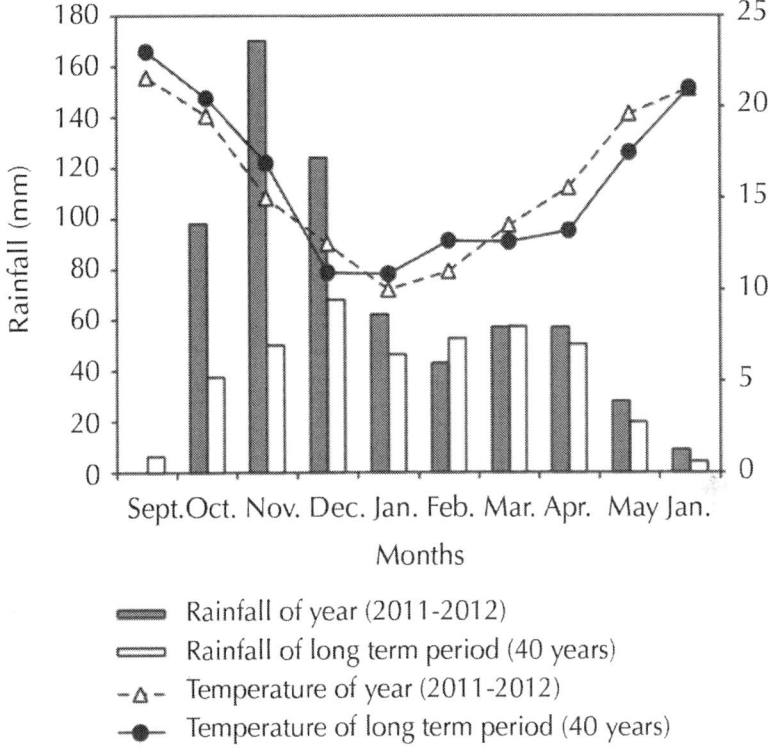

Figure 1: Monthly average of rainfall and temperature (2011-2012) at Merchouch compared to monthly values of a long term period (1970–2012).

The experiment was established in autumn 2006 after harvest of wheat. Three sites with different soil types were selected. The soils of different sites were classified as Vertisol at site 1, Cambisol at site 2, and Luvisol at site 3, according to the WRB classification [33].

Table 1: summarises the main soil characteristics of the soils at the 3 sites

Site/soil	Depth (cm)	Clay (%)	Silt (%)	Sand (%)	Texture class	pH (1.1 H2O)	P2O5 (mg/kg)	K2O (mg/kg)	CaCO3 total (g/kg)	SOC (g/kg)
Site 1 Vertisol	0–20	49	29.5	21.5	C	7.02	21	234	3.6	12.2
	20–80	52	26.6	21.4	C	7.23	11	133	7.3	11.4
	80–120	53	23.1	23.9	C	7.37	5	139	8.5	9.7
Site 2 Cambisol	0–25	44	37	18	C	7.9	58	237	1.2	11.7
	25–55	41	49	10.0	SiC	7.8	5	133	4.4	11.1
	55–90	38	55	7.4	SiL	7.7	3	108	9.5	5.5
Site 3 Luvisol	0–15	19	36	45	L	7.1	40	259	5.5	7.4
	15–50	25	28	47	L	7.2	14	129	10.5	6.5
	50–100	36	21	43	L	7.4	4	109	10.1	6.2

C: clay; SiC: silty clay; SiL: silty loam; L: loam (following the textural triangle [13])

Experimental Set-Up: In the three sites, the trials consist of two tillage systems: no-tillage system (NT) and conventional tillage system (CT), performed on two adjacent plots of 200 m long and 100 m wide each. The CT plots were ploughed, according to farmers' practice in the region, at 30 cm of depth with a "stubble plow" at the end of August each year. For seed-root bed preparation, a pass of a chisel, operated at about 15 cm depth, and two passes with a disc harrow at about 10 cm depth were needed during mid-September. The soil was not disturbed in the plots under NT which were maintained covered with flat and stubble residues at 30% levels. Wheat-lentil rotation was adopted and the crop management was similar in CT and NT treatments. Indeed, winter wheat was sown in mid-November at a 140 kg ha^{-1} seed rate. In mid-December, lentil was sown at seed rate of 40 kg ha^{-1}. Before sowing, wheat and lentil received a complex fertilizer (14N-28P$_2$O$_5$-14K$_2$O), at a rate of 150 kg ha^{1} and 100 kg ha^{-1}, respectively. In addition, wheat received 100 kg ha^{-1} of urea at the end of February. The control of weeds is based primarily on plowing the soil in CT plot before sowing; while in the NT plot this control was achieved by chemical weeding. In fact, for wheat, flumetsulam herbicide was used at a dose of 50 mL^{-1} ha before sowing. Before lentil seeding, weeds were treated with 3 l ha^{-1} of Paraquat.

NT and CT lentils were treated with 1 l ha^{-1} of fluazifop-P-butyl in early February of each season. For wheat, 50 mL ha^{-1} of flumetsulam was used in early March of each year to control weeds.

After harvest, about 30% of the crop residues were maintained at the surface under no tillage. In fact, to study the effect of residue management on SOCs under NT system, three 1 m^2 plots were randomly chosen to harvest in the NT plot in the large plot of 1 ha and the average of residues is calculated per m^2 and knowing the amount of residues collected, the mechanical combine harvester was set to leave about 30% of the residues in the large NT plot for wheat crop. For lentil, harvest was done manually and 30% of harvested residues were manually dispersed on the surface of NT plot.

In contrast with NT, in CT plots all crop residues were removed from the field according to the conventional farming practices of this region.

Soil Sampling and Analysis

Soil Organic Carbon Content

In June of 2012, after wheat harvest, disturbed and undisturbed soil samples were taken for determination of the SOC content (SOCc) and bulk density, respectively, and at the same time. The soil samples were collected in three different sites at four depths (0–5, 5–10, 10–20, and 20–30 cm) in the NT and CT plots with three replicates per treatment. We selected to use those four depths in this study because studying the effect of soil management on SOC using the entire soil profile can be more complicated and needs a long term SOC monitoring [34].

Immediately after sampling, the disturbed soil samples were dry sieved at 2 mm to remove plant debris. The SOC content was determined indirectly by oxidation of organic carbon following the classical method of Walkley and Black [35]. The organic matter content is estimated by multiplying the SOC content by a correction factor of 1.724. Concerning the soil bulk density (D_b), intact soil cores of 200 cm³ in metal sleeves were collected using a hammer-driven core sampler for the determination of dry bulk density as described by Grossman and Reinsch [36] (Table 1).

Humic Substances

To study the effect of tillage practices on humic substances, soil samples were collected randomly from the Vertisol at 0–20 cm depth in the NT and CT plots with three replicates per treatment in mid-season (March 2012).

Extraction of humic substances was done using the alkaline solvents method [37]. The fine fraction (<50 µm) was treated with 0.1 M sodium pyrophosphate at pH 9.8. From these alkaline substances, separated by centrifugation, the dispersed organomineral colloids were deflocculated with 4% KCl and mixed with insoluble soil residues. The residues were then treated with 0.1 N sodium hydroxide at pH 12 using the same protocol [38].

The pyrophosphate and NaOH extracts were mixed. The humic and fulvic acids were separated by acidification to a pH 1.5.

Calculation of Soil Organic Carbon Stock (SOCs)

The stock of soil organic carbon (SOCs) was expressed using the following equation [39]:

$$SOCs = \sum_{i=1}^{n} D_b i * SOCsi * Di, \qquad (1)$$

where SOCs is the soil organic carbon stock $(kg\,Cm^2)$, $D_b i$ is the bulk density $(Mg\,m^{-3})$ of layer , SOCci is the proportion of soil organic carbon content $(g\,C\,g^{-1})$ in layer i, and D_i is the layer depth (m).

Statistical Analysis

The effects of tillage system on the SOCc, SOCs, and humic substances were tested in the different soil types using SPSS version 17. Analysis of variance (ANOVA) was used to determine significance of tillage effects in each soil type and t-test (Student's t-test) was applied for comparing treatment means.

RESULTS AND DISCUSSION

Soil Organic Carbon Content (SOCc)

Our results showed that SOCc near the surface (0–10 cm) was significantly higher under NT compared to CT for the three soil types (Figure 2), confirming the results obtained by other authors under semiarid Mediterranean conditions [20, 32, 40, 41]. The reduction of SOCc in conventionally tilled soil could be explained by the excessive removal of biomass after harvest and higher decomposition rate due to increased microbial activity at the soil surface [42, 43]. At 10–20 cm depth, except in Luvisol, SOCc was significantly higher in NT than CT. However, in the deepest horizon (20–30 cm), no significant difference of SOCc has been observed between the CT and NT systems. The same findings under similar semiarid Mediterranean conditions were reported by Murillo et al. [44], Moreno et al. [45], and Lozano-García and Parras-Alcántara [46].

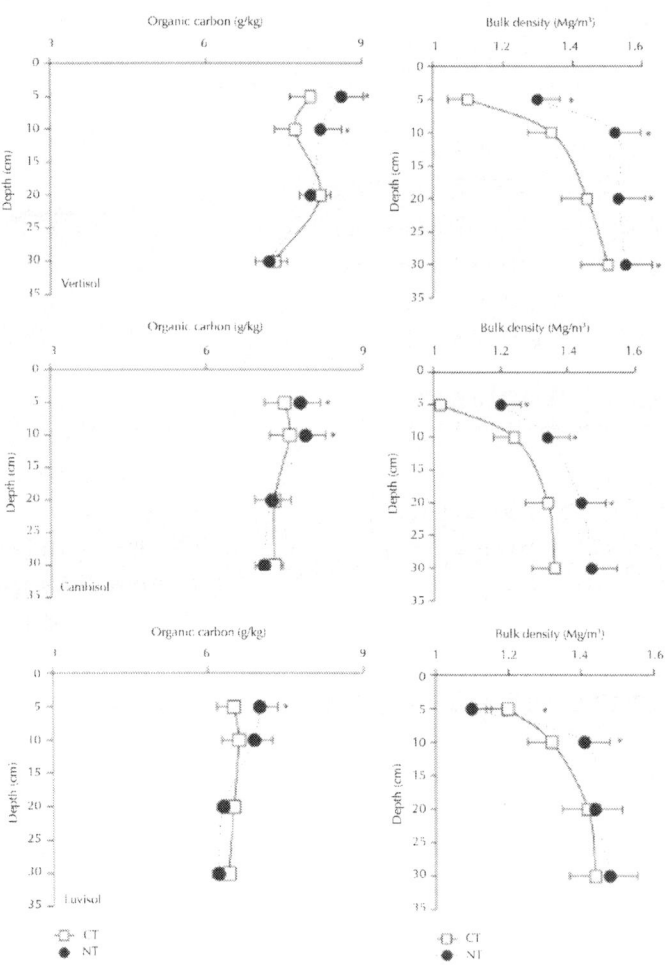

Figure 2: Soil organic carbon content (SOCc) and bulk density (D_b) in three soil types after 5 years under NT and CT. At each depth, (*) means the pres-

ence of significant differences between treatments (LSD test,p<0.05). The error bar represents one standard error.

Although under semiarid climatic conditions the potential for C sequestration into deep soil horizons is generally restricted, a slight increase of SOCc in the surface layer is essential to control erosion, water infiltration, and conservation of nutrients and is related with the soil quality [47].

Soil Organic Carbon Storage (SOCs)

Figure 2 shows higher soil D_b under NT than under CT, especially in the top soil (except for Luvisol). Similar results were reported in numerous studies, which showed that bulk density increases in the few years after NT introduction [48–50].

Several authors suggest that this increase in Db under NT will be reduced with time following increases in soil biological activity [51–53].

Table 2 shows the SOCs in all layers for both tillage systems under the three soil types. The analysis of variance indicated significant higher SOCs under NT compared to CT in both 0–5 and 5–10 cm layers under the three soil types. Considering the average SOCs at 0–30 cm depth, NT had significant higher SOCs than CT under Vertisol and Cambisol; however no significant difference was observed between the two tillage systems under Luvisol. The SOCs (0–30 cm) average was 29.35 and 27.36 Mg ha^{-1} under NT and CT, respectively, with a maximum of 31.89 Mg ha^{-1} observed in NT under Vertisol (Table 2).

Table 2: Soil organic carbon storage (SOCs) (Mg·ha^{-1}) in three sites under no tillage (NT) and conventional tillage (CT) in Merchouch, Morocco

Depth (cm)	Vertisol		Cambisol		Luvisol		Average	
	NT	CT	NT	CT	NT	CT	NT	CT
0–5	5.39a	4.14b	4.91a	4.07b	4.02a	3.31b	4.77a	3.84b
5–10	5.83a	4.62b	5.70a	4.63b	5.09a	4.69b	5.54a	4.65b
10–20	11.32a	10.77b	10.35a	9.81b	8.54a	8.61a	10.07a	9.73a
20–30	9.34a	9.26a	9.80a	9.98a	7.77a	8.18a	8.97a	9.14a
Total (0–30 cm)	31.89a	28.79b	30.76a	28.49b	25.41a	24.79a	29.35a	27.35b

a,bFor the same soil and the same depth, treatments with the same letter are not significantly different; Student's t-test (P<005).

This result indicated that after five years of continuous NT the topsoil (0–30 cm) SOCs increased by 10% in the Vertisol, 8% in the Cambisol, and 2% in Luvisol compared to CT. This is consistent with the results of Ben Moussa-Machraoui et al. [54] who observed an improvement of 12% of SOCs in a Tunisian Cambisol after 4 years of NT compared to CT.

Similarly, [18] observed a significant increase in SOCs of a Vertisol after 4 years of NT compared to CT in semiarid Morocco. In a long term study, [24] reported that SOCs in Vertisol was 13.6% higher under NT than under CT after 11 years under drier conditions compared to our study area.

In contrast, we did not observe a significant effect of NT on (0–30 cm) SOCs in the Luvisol, compared to CT. This is in agreement with results reported by Virto et al. [55] and Thomas et al. [56] in a Luvisol under semiarid climate. The higher C sequestration potential under NT of Vertisols and Cambisols versus Luvisols can be explained by differences in their texture. Several researchers have shown that fine-textured soils have a greater potential to sequester carbon than coarse-textured soils [5, 57, 58]. In fact, SOCs is higher in soils with fine texture due to the stabilizing properties that clay has on organic matter which could be trapped in the very small spaces between clay particles reducing the accessibility of the microorganisms and therefore slowing SOC decomposition. This is the mediated aggregation process by carbon accumulation as explained by Six et al. [59]. Many authors reported that soils with high clay content tend to have higher SOC than soils with low clay content under similar land use and climate conditions [27, 60].

In this sense, [61] described Vertisols as "active" soils in NT due to the high amount of clay which could improve soil organic carbon content.

Concerning the organic carbon storage, SOCs varied between 31.9 Mg·ha^{-1} and 25.8 Mg·ha^{-1} under NT, while, in tilled treatments, SOCs ranged between 28.8 Mg·ha^{-1} and 24.8 Mg·ha^{-1}. These values were lower than those observed by Fernández-Ugalde et al. [62] who found, in silty clay soil, a SOCs at 0–30 cm of 50.9 Mg·ha^{-1} after 7 years of no tillage, which was significantly higher than the 44.1 Mg·ha^{-1} under CT under wheat-barley cropping system in semiarid area in Spain. In the same Mediterranean climate, Hernanz et al. [63] found, after 11 years under NT, a SOCs of 37 Mg·ha⁻¹ which was higher than 33.5 Mg·ha⁻¹ under CT, using a wheat-vetch (Vectoria sativa L.) rotation in silty soil. The lower SOCs values we observed can be explained by the fact that more time is needed before achieving the peak sequestration rate under NT. According to West and Post [64], this peak could be reached within 5–10 years after the introduction of NT.

Humic Substances

According to Figure 3, the organic matter under NT was composed of significantly higher amounts of humic acids (HA) and humin (HU) and lower amounts of fulvic acids (FA) compared to CT in Vertisol. This is consistent with results obtained by Szajdak et al. [65]. The relative decrease in FA under NT compared to CT was probably due to the humification process which was favored by residue management under NT and resulted in a significant increase of the most stable fraction (HA and HU) [66].

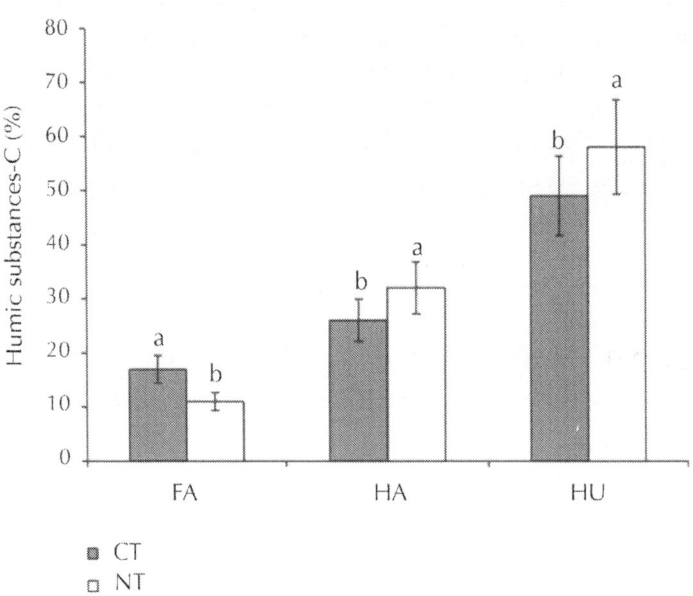

Figure 3: Humic substances in Vertisol under NT and CT (FA: fulvic acid; HA: humic acid; HU: humin), for each of the substances (treatments with the same letter are not significantly different (Student's t-test with P<0.05)).

Blanco-Canqui and Lal [67] reported that the increases of the humic acids are particularly involved in aggregate stabilization more than fulvic acid. According to Piccolo et al. [68], additional HA serve to bond particles together and can be adsorbed onto clay particles by polyvalent cations, making them especially effective in reducing clay dispersion, improving soil water content, and reducing soil erosion [69]. In fact, in the studied Vertisol, positive effects of NT on aggregate stability and soil water content have been reported in a previous study by Moussadek et al. [25].

CONCLUSIONS

The introduction of no tillage (NT) in wheat-based systems of Central Morocco could be an alternative for improving soil quality. After five years of NT, a significant increase in soil organic carbon (SOC) was found compared to CT for two major soils (Vertisol and Cambisol). SOC stock increased from 2% to 10%, depending on the soil type. Indeed, Vertisols or similar clay soils in general are able to store more SOC under NT along the profile than coarse textured soil. This accumulation of organic carbon resulted in increased levels of humic acids and humin in Vertisol under NT compared to CT. The obtained result shows that NT can contribute to the improvement of soil quality in semiarid Mediterranean conditions.

ACKNOWLEDGMENTS

The authors gratefully acknowledge the financial support of ICARDA (INRM project) and the support of Ghent University (Belgium).

REFERENCES

1. R. Lal, "Enhancing crop yields in the developing countries through restoration of the soil organic carbon pool in agricultural lands," Land Degradation and Development, vol. 17, no. 2, pp. 197–209, 2006.

2.   J. Lieskovský and P. Kenderessy, "Modelling the effect of vegetation cover and different tillage practices on soil erosion in vineyards: a case study in vráble (Slovakia) using watem/sedem," Land Degradation and Development, vol. 25, pp. 288–296, 2014.

3.   M. S. Swaminathan, "Science in response to basic human needs," Science, vol. 287, no. 5452, p. 425, 2000.

4.   P. A. Sanchez and M. S. Swaminathan, "Cutting world hunger in half," Science, vol. 307, no. 5708, pp. 357–359, 2005.

5.   R. Lal, "Soil carbon sequestration to mitigate climate change," Geoderma, vol. 123, no. 1-2, pp. 1–22, 2004.

6.   V. Barbera, I. Poma, L. Gristina, A. Novara, and M. Egli, "Long-term cropping systems and tillage management effects on soil organic carbon stock and steady state level of C sequestration rates in a semiarid environment," Land Degradation and Development, vol. 23, no. 1, pp. 82–91, 2012.

7.   F. García-Orenes, C. Guerrero, A. Roldán et al., "Soil microbial biomass and activity under different agricultural management systems in a semiarid Mediterranean agroecosystem," Soil and Tillage Research, vol. 109, no. 2, pp. 110–115, 2010.

8.   A. Bot and J. Benites, "The importance soil organic matter: key to drought-resistant soil sustained food production," FAO Soils Bulletin, vol. 80, 2005.

9.   B. Lozano-García and L. Parras-Alcántara, "Variation in soil organic carbon and nitrogen stocks along a toposequence in a traditional Mediterranean olive grove," Land Degradation and Development, vol. 25, pp. 297–304, 2014.

10.   C. H. Srinivasarao, B. Venkateswarlu, R. Lal et al., "Long-term manuring and fertilizer effects on depletion of soil organic stocks under Pearl millet-cluster vean -castor rotation in Western india," Land Degradation and Development, vol. 25, no. 2, pp. 173–183, 2014.

11.   B. Soudi, C. N. Chiang, H. Berdai, and F. Naaman, "Statut du cycle de lazote et de la matière organique en zones semi-arides irriguées et dagriculture pluviale," Revue H.T.E 127, 2003.

12.   R. Mrabet, K. Ibno-Namr, F. Bessam, and N. Saber, "Soil chemical quality changes and implications for fertilizer management after 11 years of no-tillage wheat production systems in Semiarid

Morocco," Land Degradation and Development, vol. 12, no. 6, pp. 505–517, 2001.

13. Soil Conservation Service, vol. 18, Survey Staff, Soil Survey Manual, U.S. Department of Agriculture Handbook, 1993.

14. R. Derpsch and T. Friedrich, "Global overview of conservation agriculture adoption," in Proceedings of the 4th World Congress on Conservation Agriculture, pp. 429–438, New Delhi, India, February 2009.

15. G. Zhao, X. Mu, Z. Wen, F. Wang, and P. Gao, "Soil erosion, conservation, and Eco-environment changes in the Loess Plateau of China," Land Degradation & Development, vol. 24, pp. 499–510, 2013.

16. A. Tesfaye, W. Negatu, R. Brouwer, and P. Van der Zaag, "Understanding soil conservation decision of farmers in the gedeb watershed, ethiopia," Land Degradation and Development, vol. 25, pp. 71–79, 2014.

17. A. Kassam, T. Friedrich, R. Derpsch et al., "Conservation agriculture in the dry Mediterranean climate,"Field Crops Research, vol. 132, pp. 7–17, 2012.

18. R. Mrabet, "Mediterranean conservation agriculture. A paradigm for cropping systems," in Proceedings of the 16th Triennial Conference of International Soil Tillage Research Organization (ISTRO '03), pp. 774–779, Brisbane, Australia, July 2003.

19. R. Mrabet, "Climate change and carbon sequestration in the Mediterranean basin. Contributions of no-tillage systems," Options Méditerranéennes, vol. 96, pp. 165–184, 2010.

20. J. Álvaro-Fuentes, M. V. López, C. Cantero-Martinez, and J. L. Arrúe, "Tillage effects on soil organic carbon fractions in Mediterranean dryland agroecosystems," Soil Science Society of America Journal, vol. 72, no. 2, pp. 541–547, 2008.

21. F. Basso, M. Pissante, and B. Basso, "Soil erosion and land degradation," in Mediterranean Desertification. A Mosaic of Processes and Responses, N. A. Geeson, C. J. Brandt, and J. B. Thornes, Eds., pp. 347–359, John Wiley & Sons, New York, NY, USA, 2002.

22. K. Oorts, H. Bossuyt, J. Labreuche, R. Merckx, and B. Nicolardot, "Carbon and nitrogen stocks in relation to organic matter

fractions, aggregation and pore size distribution in no-tillage and conventional tillage in northern France," European Journal of Soil Science, vol. 58, no. 1, pp. 248–259, 2007.

23. R. Mrabet, Crop residue management and tillage systems for water conservation in a semiarid area of Morocco [Ph.D. dissertation], Colorado State University, Fort Collins, Colo, USA, 1997.

24. R. Mrabet, N. Saber, A. El-Brahli, S. Lahlou, and F. Bessam, "Total, particulate organic matter and structural stability of a Calcixeroll soil under different wheat rotations and tillage systems in a semiarid area of Morocco," Soil and Tillage Research, vol. 57, no. 4, pp. 225–235, 2001.

25. R. Moussadek, R. Mrabet, P. Zante et al., "Influence du semis direct et des résidus de culture sur l'érosion hydrique d'un Vertisol Méditerranéen," Canadian Journal of Soil Science, vol. 91, no. 4, pp. 627–635, 2011.

26. R. Mrabet, "Soil quality and carbon sequestration. Impacts of no-tillage systems," Options Méditerranéennes, vol. 65, pp. 45–60, 2006.

27. A. J. Franzluebbers, "Tillage and residue management effect on soil organic matter," in Soil Organic Matter in Sustainable Agriculture, F. Magdoff and R. R. Weil, Eds., pp. 227–268, CRC Press, Boca Raton, Fla, USA, 2004.

28. K. H. Tan, Environmental Soil Science, Marcel Dekker, New York, NY, USA, 1994.

29. G. Ding, J. M. Novak, D. Amarasiriwardena, P. G. Hunt, and B. Xing, "Soil organic matter characteristics as affected by tillage management," Soil Science Society of America Journal, vol. 66, no. 2, pp. 421–429, 2002.

30. D. L. McCallister and W. L. Chien, "Organic carbon quantity and forms as influenced by tillage and cropping sequence," Communications in Soil Science and Plant Analysis, vol. 31, no. 3-4, pp. 465–479, 2000.

31. R. Madrid, M. Valverde, I. Guillén, A. Sanchez, and A. Lax, "Evolution of organic matter added to soils under cultivation conditions," Journal of Plant Nutrition and Soil Science, vol. 167, no. 1, pp. 39–44, 2004.

32. R. Mrabet, No-Tillage Systems for Sustainable Dry Land Agriculture in Morocco, INRA Publication, Fanigraph edition, 2008.

33. WRB, World Reference Base for Soil Resources, World Soil Resources Reports No. 103, FAO, Rome, Italy, 2nd edition, 2006.

34. L. Parras-Alcántara, M. Martín-Carrillo, and B. Lozano- García, "Impacts of land use change in soil carbon and nitrogen in a Mediterranean agricultural area (Southern Spain)," Solid Earth, vol. 4, pp. 167–177, 2013.

35. A. Walkley and I. A. Black, "An examination of Degtjareff method for determining soil organic matter, and a proposed modification of the chromic acid titration method," Soil Science, vol. 37, no. 1, pp. 29–38, 1934.

36. R. B. Grossman and T. G. Reinsch, "Bulk density and linear extensibility," in Methods of Soil Analysis: Part 4, Physical Methods, J. H. Dane and G. C. Topp, Eds., SSSA Book Series, pp. 201–228, Soil Science Society of America, Madison, Wis, USA, 2002.

37. A. Piccolo, "Characteristics of soil humic matter addition by caulescent Andean rosettes on surficial soil," Geoderma, vol. 54, pp. 151–171, 1988.

38. B. Dabin, "Méthode d'extraction et de fractionnement des matières humiques du sol: application à quelques études pédologiques et agronomiques dans les sols tropicaux," Cahiers ORSTOM, Série Pédologie, vol. 14, pp. 287–297, 1976.

39. M. Bernoux, M. D. C. S. Carvalho, B. Volkoff, and C. C. Cerri, "Brazil's soil carbon stocks," Soil Science Society of America Journal, vol. 66, no. 3, pp. 888–896, 2002.

40. P. Bescansa, M. J. Imaz, I. Virto, A. Enrique, and W. B. Hoogmoed, "Soil water retention as affected by tillage and residue management in semiarid Spain," Soil and Tillage Research, vol. 87, no. 1, pp. 19–27, 2006.

41. L. López-Bellido, F. J. López-Garrido, M. Fuentes, J. E. Castillo, and E. J. Fernández, "Influence of tillage, crop rotation and nitrogen fertilization on soil organic matter and nitrogen under rain-fed Mediterranean conditions," Soil and Tillage Research, vol. 43, no. 3-4, pp. 277–293, 1997.

42. D. L. Karlen, N. C. Wollenhaupt, D. C. Erbach et al., "Crop residue effects on soil quality following 10-years of no-till corn," Soil and Tillage Research, vol. 31, no. 2-3, pp. 149–167, 1994.

43. R. A. Drijber, J. W. Doran, A. M. Parkhurst, and D. J. Lyon, "Changes in soil microbial community structure with tillage under long-term wheat-fallow management," Soil Biology and Biochemistry, vol. 32, no. 10, pp. 1419–1430, 2000.

44. J. M. Murillo, F. Moreno, F. Pelegrín, and J. E. Fernández, "Responses of sunflower to traditional and conservation tillage under rainfed conditions in southern Spain," Soil and Tillage Research, vol. 49, no. 3, pp. 233–241, 1998.

45. F. Moreno, J. M. Murillo, F. Pelegrín, and I. F. Girón, "Long-term impact of conservation tillage on stratification ratio of soil organic carbon and loss of total and active $CaCO_3$," Soil and Tillage Research, vol. 85, no. 1-2, pp. 86–93, 2006.

46. B. Lozano-García and L. Parras-Alcántara, "Land use and management effects on carbon and nitrogen in Mediterranean Cambisols," Agriculture, Ecosystems & Environment, vol. 179, pp. 208–214, 2013.

47. A. J. Franzluebbers, F. M. Hons, and D. A. Zuberer, "Tillage and crop effects on seasonal dynamics of soil CO_2 evolution, water content, temperature, and bulk density," Applied Soil Ecology, vol. 2, no. 2, pp. 95–109, 1995.

48. D. K. Cassel, C. W. Raczkowski, and H. P. Denton, "Tillage effects on corn production and soil physical conditions," Soil Science Society of America Journal, vol. 59, no. 5, pp. 1436–1443, 1995.

49. A. J. Franzluebbers, "Soil organic matter stratification ratio as an indicator of soil quality," Soil and Tillage Research, vol. 66, no. 2, pp. 95–106, 2002.

50. J. Lampurlanés and C. Cantero-Martínez, "Soil bulk density and penetration resistance under different tillage and crop management systems and their relationship with barley root growth," Agronomy Journal, vol. 95, no. 3, pp. 526–536, 2003.

51. L. R. Drees, A. D. Karathanasis, L. P. Wilding, and R. L. Blevins, "Micromorphological characteristics of long-term no-till and conventionally tilled soils," Soil Science Society of America Journal, vol. 58, no. 2, pp. 508–517, 1994.

52. A. Jordán, L. M. Zavala, and J. Gil, "Effects of mulching on soil physical properties and runoff under semi-arid conditions in southern Spain," Catena, vol. 81, no. 1, pp. 77–85, 2010.

53. R. Mrabet, R. Moussadek, A. Fadlaoui, and E. van Ranst, "Conservation agriculture in dry areas of Morocco," Field Crops Research, vol. 132, pp. 84–94, 2012.

54. S. Ben Moussa-Machraoui, F. Errouissi, M. Ben-Hammouda, and S. Nouira, "Comparative effects of conventional and no-tillage management on some soil properties under Mediterranean semi-arid conditions in northwestern Tunisia," Soil and Tillage Research, vol. 106, no. 2, pp. 247–253, 2010.

55. I. Virto, M. J. Imaz, A. Enrique, W. Hoogmoed, and P. Bescansa, "Burning crop residues under no-till in semi-arid land, Northern Spain: effects on soil organic matter, aggregation, and earthworm populations," Australian Journal of Soil Research, vol. 45, no. 6, pp. 414–421, 2007.

56. G. A. Thomas, R. C. Dalal, and J. Standley, "No-till effects on organic matter, pH, cation exchange capacity and nutrient distribution in a Luvisol in the semi-arid subtropics," Soil and Tillage Research, vol. 94, no. 2, pp. 295–304, 2007.

57. J. Hassink, "Decomposition rate constants of size and density fractions of soil organic matter," Soil Science Society of America Journal, vol. 59, no. 6, pp. 1631–1635, 1995.

58. B. A. Needelman, M. M. Wander, G. A. Bollero, C. W. Boast, G. K. Sims, and D. G. Bullock, "Interaction of tillage and soil texture: biological active soil organic matter in Illinois," Soil Science Society of America Journal, vol. 63, no. 5, pp. 1326–1334, 1999.

59. J. Six, C. Feller, K. Denef, S. M. Ogle, J. C. de Moraes Sa, and A. Albrecht, "Soil organic matter, biota and aggregation in temperate and tropical soils—effects of no-tillage," Agronomie, vol. 22, no. 7-8, pp. 755–775, 2002.

60. L. Parras-Alcántara, L. Díaz-Jaimes, and B. Lozano-García, "Management effects on soil organic carbon stock in Mediterranean open rangelands—treeless grasslands," Land Degradation & Development, 2014.

61. R. Lal, "Conservation tillage for sustainablc agriculture," Advances in Agronomy, vol. 42, pp. 185–197, 1988.

62. O. Fernández-Ugalde, I. Virto, P. Bescansa, M. J. Imaz, A. Enrique, and D. L. Karlen, "No-tillage improvement of soil physical quality in calcareous, degradation-prone, semiarid soils," Soil and Tillage Research, vol. 106, no. 1, pp. 29–35, 2009.

63. J. L. Hernanz, V. Sánchez-Girón, and L. Navarrete, "Soil carbon sequestration and stratification in a cereal/leguminous crop rotation with three tillage systems in semiarid conditions," Agriculture, Ecosystems and Environment, vol. 133, no. 1-2, pp. 114–122, 2009.

64. T. O. West and G. Post, "Soil organic carbon sequestration rates by tillage and crop rotation: a global data analysis," Soil Science Society of America Journal, vol. 66, no. 6, pp. 1930–1946, 2002.

65. L. Szajdak, A. Jezierski, and M. L. Cabrera, "Impact of conventional and no-tillage management on soil amino acids, stable and transient radicals and properties of humic and fulvic acids," Organic Geochemistry, vol. 34, no. 5, pp. 693–700, 2003.

66. M. G. González, M. E. Conti, R. M. Palma, and N. M. Arrigo, "Dynamics of humic fractions and microbial activity under no-tillage or reduced tillage, as compared with native pasture (Pampa Argentina)," Biology and Fertility of Soils, vol. 39, no. 2, pp. 135–138, 2003.

67. H. Blanco-Canqui and R. Lal, "Mechanisms of carbon sequestration in soil aggregates," Critical Reviews in Plant Sciences, vol. 23, no. 6, pp. 481–504, 2004.

68. A. Piccolo, G. Pietramellara, and J. S. C. Mbagwu, "Use of humic substances as soil conditioners to increase aggregate stability," Geoderma, vol. 75, no. 3-4, pp. 267–277, 1997.

69. J. Balesdent, "Un point sur l'évolution des réserves organiques des sols de France," Etude et Gestion des Sols, vol. 3, no. 4, pp. 245–260, 1996.

Chapter 8

Soil Quality Assessment Strategies for Evaluating Soil Degradation in Northern Ethiopia

Gebreyesus Brhane Tesfahunegn[1, 2]

[1]College of Agriculture, Aksum University-Shire Campus, 314 Shire, Ethiopia
[2]Center for Development Research (ZEF), University of Bonn, Walter-Flex Street No. 3, 53113 Bonn, Germany

ABSTRACT

Soil quality (SQ) degradation continues to challenge sustainable development throughout the world. One reason is that degradation indicators such as soil quality index (SQI) are neither well documented nor used to evaluate current land use and soil management systems (LUSMS). The objective was to assess and identify an effective SQ indicator dataset from among 25 soil measurements, appropriate scoring functions for each indicator and an efficient SQ indexing method to evaluate soil degradation across the LUSMS in the Mai-Negus catchment of northern Ethiopia. Eight LUSMS selected for soil sampling and analysis included (i) natural forest (LS1), (ii) plantation of protected area, (iii) grazed land, (iv) teff (Eragrostis tef)-faba bean (Vicia faba) rotation, (v) teff-wheat (Triticum vulgare)/barley (Hordeum vulgare) rotation, (vi) teff monocropping, (vii) maize (Zea mays) monocropping, and (viii) uncultivated marginal land (LS8). Four principal components explained almost 88% of the variability among the LUSMS. LS1 had the highest mean SQI (0.931) using the scoring functions and principal component analysis (PCA) dataset selection, while the lowest SQI (0.458) was measured for LS8. Mean SQI values for LS1 and LS8 using expert opinion dataset selection method were 0.874 and 0.406, respectively. Finally, a sensitivity analysis (S) used to compare PCA and expert opinion dataset selection procedures for various scoring functions ranged from 1.70 for unscreened-SQI to 2.63 for PCA-SQI. Therefore, this study concludes that a PCA-based SQI would be the best way to distinguish among LUSMS since it appears more sensitive to disturbances and management practices and could thus help prevent further SQ degradation.

INTRODUCTION

Globally, declining in soil quality (SQ) has posed a tremendous challenge to increasing agricultural productivity, economic growth, and healthy environment [1, 2]. The underlying causes for SQ degradation are largely related to inappropriate land use and soil management, erratic and erosive rainfall, steep terrain, deforestation, and overgrazing [2–4]. Most of the causes are resulted from a desperate attempt by farmers to increase production for the growing population which aggravate SQ degradation more in the developing countries, which mainly depend on natural resources (agriculture) [1, 4]. Misuse of natural resources that leads to degradation can also be stimulated by socioeconomic and political issues, for example, land tenure, capital, and infrastructure [5]. SQ degradation by soil erosion such as soil nutrient depletion and changes in soil physical indicators is largely recognized as a principal cause aggravated by the effect of inappropriate land use and soil management in the developing countries like Ethiopia [6, 7].

Interest in the evaluation of soil degradation particularly the quality of soil resources has been increased as soil is critically important component of the Earth's biosphere, functioning not only in the production of food and fibers but also in the maintenance of environmental quality [8]. In normal conditions, the soil can maintain equilibrium by pedogenetic processes [9–11]. However, this equilibrium is easily disturbed by anthropogenic activities (e.g., agricultural practices, deforestation, and overgrazing), and such effects are mainly noticed in the developing countries with poor technical and financial resources to manage natural resources [10, 11]. In order to make sound decisions regarding sustainable land use systems, knowledge of SQ related to different land use scenarios is essential [12]. It is therefore most important to assess SQ degradation of different land use and soil management systems using soil quality index (SQI) since many of the factors that influence sustainable productivity are related to SQ. Information on SQI can support to further prioritization and then device management strategies that improve soil resources sustainably [11]. To do so, applying the concept of SQI is desirable as individual soil properties in isolation may not be sufficient to quantify changes in SQ related to land use and soil management systems [11, 13]. In line to this, many studies reported that indexing SQ indicators based on a combination of soil properties could better reflect the status of SQ degradation as compared to individual parameters [6, 13–15].

Despite the importance of SQI in describing SQ degradation or aggradations, there is no universally accepted dataset selection, scoring, and SQ indexing method for field conditions. Previous studies reported that different methods of minimum dataset selection (MDS), scoring, and SQ indexing have been applied but SQI results varied even for the same conditions [11, 13, 16]. The most widely reported MDS methods of SQ indicators are expert opinion and statistical tools (e.g., regression, principal component analysis (PCA)) [11, 13]. An expert can generate a list of appropriate SQ indicators on the basis of ecosystem processes and functions and other decision rules such as management goals for a site associated with soil functions as well as other site-specific factors, like region or crop sensitivity as selection criteria [13, 16].

The transformation of the datasets into scores (scoring function) can be done using linear and nonlinear scoring techniques [11, 13]. Studies elsewhere compared the two scoring methods to represent soil system function but the value of nonlinear scoring method was reported higher than the linear method [11, 16, 17] There are different types of linear and nonlinear scoring functions, even though none of the previous studies have evaluated them all simultaneously [11, 16]. Different SQ indexing methods have been also used by different researchers [13, 16–18]. The same authors have reported that there are differences in SQI values among the various SQ indexing methods (e.g., additive, weighted, and max-min objective functions). Despite the fact that there is diversity in data selection, scoring, and SQ indexing methods, previous studies have limitation in evaluating the methods using the same data simultaneously in a similar field conditions.

Regardless of the above limitation, having SQI of long-term land use and soil management systems is necessary in order to locate areas to be carefully managed for sustainable development. The use of site-specific SQI can help planners and decision makers to evaluate which land use and management system is most sustainable and vice-versa in a given situation [18, 19]. These authors also noted that SQI can reflect the extent of SQ degradation and thereby give support to suggest appropriate remedial measures such as optimum fertilizer rates and planning of other suitable land management practices considering potentials and constraints of different fields at large scale such as a catchment.

In general, SQI is a useful assessment tool that may help move soil conservation and resource management beyond assessments of soil erosion and changes in productivity [13]. SQI can thus provide the necessary information for planners and decision makers to make informed decisions against SQ degradation using the introduction of appropriate interventions. Despite such importance of SQI in combating SQ degradation, only few studies have been reported in relation to various land use and soil management systems. This indicated that research on SQI has been mostly neglected for unknown reasons, with the most probable reason which could be technical and financial limitations.

Many approaches assessing SQ degradation using the concept of SQI have been already developed and applied elsewhere [6, 11, 13, 15–19]. In this study, such concepts are adopted and evaluated to narrow the knowledge/information gap of SQI across different land use and soil management systems in the northern Ethiopia. The objective of this study was to assess and identify an effective SQ indicator dataset among 25 soil measurements, appropriate scoring functions for each indicator, and an efficient SQ indexing method to evaluate soil degradation across the LUSMS in the Mai-Negus catchment of northern Ethiopia.

MATERIALS AND METHODS

Study Area

This evaluation was conducted in the Mai-Negus catchment in Tigray regional state, northern Ethiopia (Figure1). The catchment covers an area of 1240 ha, with a landscape consisting of generally rugged terrain at altitudes ranging from 2060 to 2650 m above sea level. Land use is dominantly arable with teff (Eragrostis tef) being the primary crop on >80% of the land area. The remainder of the catchment is pasture with scattered patches of mixed tree, bush, and shrub cover. The major rock types are lava pyroclastic and metavolcanic. According to FAO-UNESCO Soil Classification System, soils are dominantly Leptosols at very steep positions, Cambisols on middle to steep slopes, and Vertisols on flat areas [20]. Annual rainfall averages 700 mm but is very erratic in amount and distribution throughout the catchment. Mean annual temperature was 22°C.

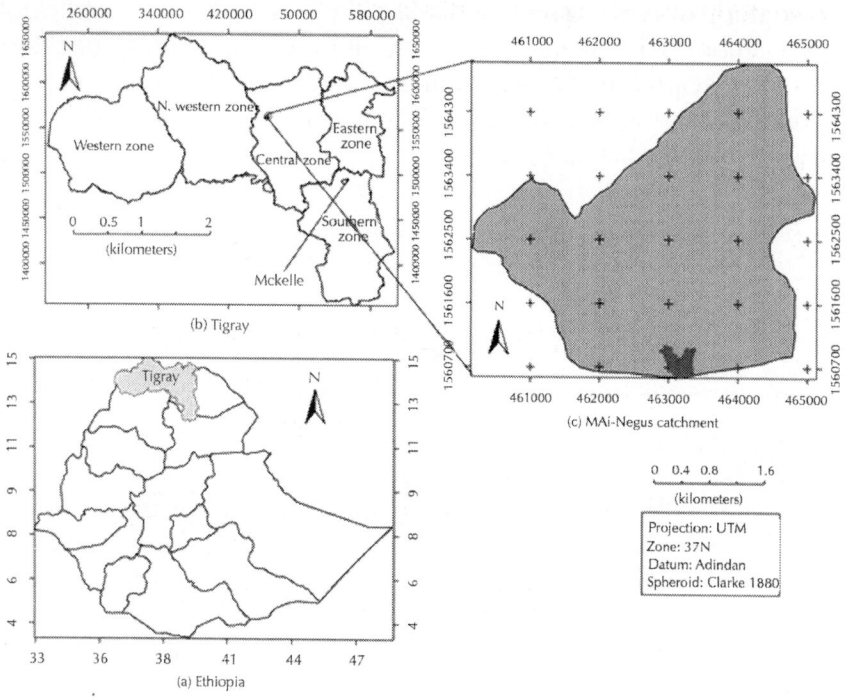

Figure 1: Locations of the study area: Ethiopia (a), Tigray Region (b), and Mai-Negus catchment (study site) (c).

Selection of Land Use and Soil Management Systems

Eight land use and soil management systems (LUSMS) were selected on the basis of three steps. First, information on historical and existing LUSMS in the catchment was collected and described. Soil sampling units were then identified across each LUSMS. Finally, composite soil samples were collected, processed, and analyzed for several SQ indicators using standard laboratory procedures. The first step is described below, with details for the second and third steps given in Sections 2.3 and 2.4.

Field reconnaissance surveys and informal group discussions were conducted in June 2009 by the author, two development agents, and six farmers who are knowledgeable about the catchment and local farming systems. The six farmers were selected purposively because the large group size made it impractical for all the household heads to participate and doing so would have been problematic for discussion and consensus building. The dominant cropping history and soil management practices for each LUSMS were identified and described by the team. In addition, terrain characteristic and soil factors were documented for each rain-fed agricultural LUSMS. All eight LUSMS were selected as much as possible to be from similar soil type (Cambisols) and a range in slope gradient in the catchment. Topographical characteristics of each sampling units are presented in Table 1.

Table 1: Topographic characteristics of each sampling unit in the eight LUSMS selected for soil quality assessment within the Mai-Negus catchment in northern Ethiopia

LUSMS	Sampling Unit 1				Sampling Unit 2				Sampling Unit 3			
	Elevation (m)	Slope (%)	UTMa Latitude	Longitude	Elevation (m)	Slope (%)	UTM Latitude	Longitude	Elevation (m)	Slope (%)	UTM Latitude	Longitude
LS1	2175	7	463657	1561482	2170	5.5	463856	1561592	2173	6.2	464039	1561523
LS2	2124	4.5	463383	1560411	2152	5	463615	1560503	2138	5.5	463248	1560503
LS3	2136	5	463425	1561193	2165	5.5	463489	1561466	2151	5.0	463535	1561614
LS4	2168	6	463592	1561250	2145	6.5	462538	1561454	2157	6.0	463168	1561832
LS5	2141	6.5	463606	1561338	2163	7	463317	1561763	2152	6.8	463707	1561912
LS6	2165	6	463734	1561459	2136	5.5	463363	1561500	2151	5.8	464124	1561328
LS7	2162	6.5	462810	1561314	2155	7	462641	1561614	2159	6.8	464027	1561202
LS8	2149	7	462795	1561152	2138	5	462745	1561718	2144	6.0	463913	1562153

[a]Universal Transverse Mercator 37 North (UTM-37N) in meters is the projection system.

LUSMS: land use and soil management systems; LS1: natural forest (reference); LS2: plantation of protected area; LS3: grazed land; LS4: teff (Eragrostis tef)-faba bean (Vicia faba) rotation; LS5: teff-wheat (Triticum vulgare)/Barley (Hordeum vulgare) rotation; LS6: teff monocropping; LS7: Maize (Zea mays) mono-cropping; LS8: uncultivated-marginal land system.

Based on land use information acquired in the study, eight LUSMS that represent the best and worst management practices being used throughout the study catchment were identified and are described (Table 2). The LUSMS selected for SQ evaluation were (i) natural forest (LS1), (ii) plantation on protected areas (LS2), (iii) grazed land (LS3), (iv) teff (Eragrostis tef)-faba bean (Vicia faba) rotations (LS4), (v) teff-wheat (Triticum vulgare)/barley (Hordeum vulgare) rotations (LS5), (vi) teff monocropping (LS6), (vii) maize (Zea mays)monocropping (LS7), and (viii) uncultivated marginal land (LS8). The various LUSMS were in place for various amounts of time ranging from 5 to 6 years for teff monocropping and 20 to 30 years for maize monocropping. Average age for the other systems was about 10 years except for the plantation, grazed land, and uncultivated marginal land with which was in place for more than 15 years. To assess the impact of LUSM on SQ indicators, it is either necessary to have a baseline against which human induced differences can be measured [21] or to measure the same systems repeatedly in time [13]. For this study, the researcher chose to use the natural forest (LS1) as a reference, assuming the soil in those areas is less disturbed than in cultivated or grazed areas.

Table 2: Description of the eight land use and soil management systems (LUSMS) evaluated within the Mai-Negus catchment in northern Ethiopia

Serial number	Land use and soil management systems	Description
1	Natural forest (LS1)	Less disturbed land, used as a reference in the system, which has native trees, vegetation, and grass cover.
2	Plantation of protected area (LS2)	Sesbania (Sesbania sesban) and Leucaena (Leucaena leucocephala) trees plantation was established 16 years ago and grass was used by cuting and carrying during the dry season for livestock, protected throughout the year from livestock interferences; no fertilizer application and less intensive soil conservation measures exist.

3	Grazed land (LS3)	Open grazed land appeared during the dry season 16 years ago with no inclusion of any improved management practices, for example, soil and water conservation and enrichment of plant/grass species. It overstocked in the dry months (November–June) but it is a swampy area for the rest of months.
4	Teff (Eragrostis tef)-faba bean (Vicia faba) rotation (LS4)	Fields were harvested of teff (Eragrostic tef (Zucc) Trot) crop before soil samples were collected and rotated with faba bean (Vicia faba L.) for more than 5-6 years. Urea and diammonium phosphate (DAP) fertilizers were applied each for teff at 50 kg ha−1 y−1 but sometimes reduced by half depending on resource availability and quality of the soil. Teff needed 4–6 times tillage and at least one time weeding. For faba bean, 2-3 times tillage was sufficient with no or one time weeding and addition of manure around homestead is common practice but is urea or DAP rarely used. Soil and water conservation was used at field borders.
5	Teff-wheat (Triticum vulgare)/Barley (Hordeum vulgare) rotation (LS5)	The fields were planted wheat (Tritium vulgare L.) before teff (Eragrostic tef (Zucc) Trot) and soil samples were collected after teff was harvested. Wheat (Triticum vulgare L.)/barley (Hordeum vulgare L.) with teff (Eragrostic tef) was rotated for more than 6 years. 50 kg ha−1 y−1 of each urea and DAP fertilizers was used for teff field but the amount varies with crop color at vegetative stage and soil quality condition for fertilizer rate application on wheat/barley crop fields. The fertilizer rate used for wheat/barley is lower than teff.
6	aTeff mono-cropping land system (LS6)	For more than 5-6 years, teff (Eragrostic tef (Zucc) Trot) was sown continuously with 50 kg ha−1 y−1 urea and 100 kg ha−1 y−1 DAP fertilizers; 5–7 tillage frequency; at least one time hand weeding. Manure and intensive SWC for fields around homestead was applied.

| 7 | Maize (Zea mays L.)-mono cropping land system (LS7) | Maize (Zea mays L.) was planted for more than 20–30 years continuously. The addition of manure was estimated at 6–12 t ha−1 y−1, which varies with manure and labor availability for transportation; 2-3 tillage frequency with at least one times hand hoeing and weeding. This is always practiced around homestead fields. Conservation measures are also well executed. |
| 8 | Uncultivated marginal land system (LS8) | It was terraced with wide spacing since the last 20 years but most of it is broken and used as an open grazing land throughout the year; on some spot areas it had very few to few naturally growing but over grazed grass species like Bermuda grass (Cynodon dactylon L.) management practices such as fertilizer and species enrichment were not introduced. Farmers considered it as the most degraded soil, for example, abandoned land. |

[a]Teff is the dominant crop in the study catchment and other parts of northern Ethiopia. It is an annual cereal crop (belonging to the grass family) which has sparse crop canopies and provides little cover to the soil against erosion. It has very fine seeds that require repeated plowing of fields and preparation of fine seedbeds, which increases the vulnerability of the soil to erosion.

Soil Sampling, Processing, and Analysis

After identifying the eight LUSMS locations, three soil sampling units and their corresponding areas were selected by the researcher considering representativeness and uniformity of the fields. Soil samples were collected from 24 areas (8 LUSMS × 3 sampling units) in June 2009 and analyzed for 24 potential SQ indicators. Each soil sampling unit ranged from 50 to 80 m². Five to eight soil samples from 0 to 20 cm depth (plow layer) were collected randomly from each unit and mixed to form a composite sample. The number of composite samples was determined by the size and homogeneity (hydrologic conditions) of each sampling unit. Fewer samples were collected from homogenous, small fields compared to large, heterogeneous fields. The sampling focused on the plow layer because this is where most SQ changes are expected to occur due to long-term land use and soil management practices. Each composite soil sample was mixed thoroughly in a bucket before taking a 500 g subsample that was air dried and sieved to pass a 2 mm mesh before analysis.

The soil samples were analyzed for selected physical, chemical, and biological SQ indicators. Soil texture was determined using the Bouyoucos hydrometer method [22] and soil bulk density (BD) by the core method [23]. Percent pore space (total porosity) was computed from BD and average particle density (PD) of $2.65\,g\,cm^{-3}$ as [24]. Soil aggregate stability (SAS) was measured using the wet sieve method [25] and maximum water holding capacity (MWHC) determined by equilibrating the soil with water through capillary action in a KR box [26]. A-horizon depth was directly measured in pits opened to a 60 cm depth within each LUSMS.

Soil pH was determined using a 1:2.5 soil to water ratio with a combined glass electrode [27]. Soil organic (OC) was determined by the Walkley-Black method [28], available phosphorus (Pav) by the Olsen method [29], total nitrogen (TN), and total phosphorus (TP) by the Kjeldahl digestion method [30]. Cation exchange capacity (CEC) was determined by ammonium acetate buffered at pH 7 [31]. Exchangeable bases (calcium, Ca; magnesium, Mg; potassium, K) were analyzed after extraction using 1 M ammonium acetate at pH 7.0. Iron and zinc were determined using 0.005 M diethylene triamine pentaacetic acid (DTPA) extraction as described in Baruah and Barthakur [26].

Exchangeable sodium percentage (ESP) was calculated by dividing exchangeable Na^+ by CEC. Base saturation percentage (BSP) was calculated by dividing the sum of base forming cations by CEC, multiplied by 100% [32]. A 25th SQ indicator, earthworm population, was monitored monthly as a biological indicator throughout the wet season (mid-June–mid-September 2009). Three randomly collected soil samples ($25 \times 25 \times 20$ cm) from a 1 m² area were passed through a 10 mm sieve to separate and then count the average number of earthworms in each LUSMS sampling unit.

Soil Quality Index Computations

After measuring the 25 potential SQ indicators using field and laboratory analysis techniques, different methods for computing SQI values were evaluated. Although the type of data used for each SQI may differ, the process of SQ indexing follows the same three basic steps regardless of the method used (Figure 2). These steps are indicator selection, interpretation/scoring, and integration into index value [11, 33]. The detailed descriptions of each step are given below.

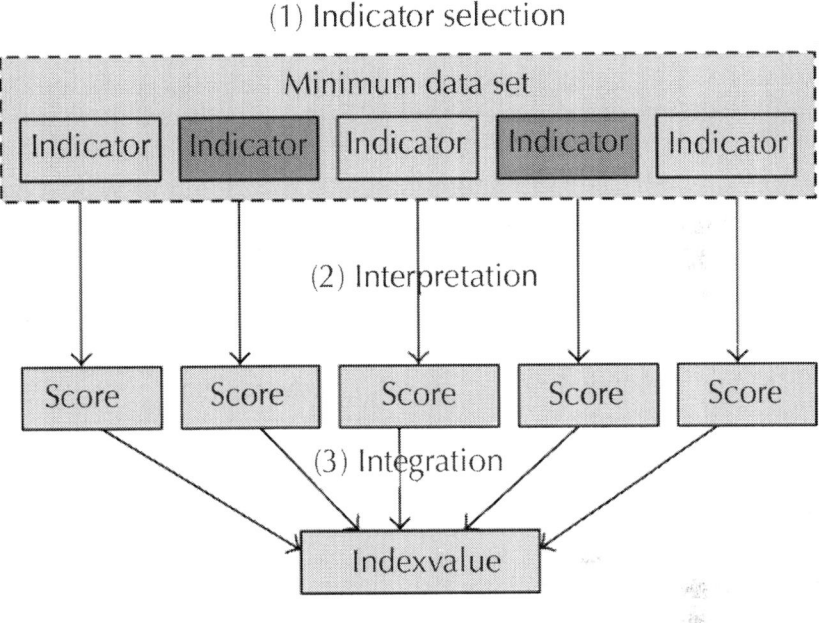

Figure 2: A conceptual model for computing soil quality indices. (From Andrews [33].

Indicator Selection (Step 1)

Potential SQ indicators were selected based on their sensitivity to management practices, ability to describe major soil processes, ease and cost of sampling and laboratory analysis, and significance of increasing productivity (agronomic) and protecting environmental soil functions. Two methods for selecting a minimum dataset (MDS), expert opinion (EO), and principal component analysis (PCA) were compared. For the EO approach MDS, variables were chosen from the 25 potential SQ indicators based on researcher knowledge and literature recommendations [34–36].

Indicator Transformation/Scoring Function (Step 2)

After selecting the MDS using EO or PCA (Step 1), each value was transformed using two different scoring techniques [11, 33]. Both linear and nonlinear scoring functions were compared as described below.

Linear Scoring Functions: Three linear scoring functions (LSF) were identified and evaluated. For the first LSF approach, indicators were ranked in ascending for "more is better" or descending order for "less is better" in terms of soil functions [37]. Each "more is better" indicator was divided by the highest value in the group such that it received a score of 1. For "less is better" indicator with the lowest observed value, it was divided by itself so that it received a score of 1. Threshold values were identified and used as outlined by Liebig et al. [37]. This approach (I^L) was therefore described as the "Liebig Linear Scoring Function."

For the second LSF approach, SQ indicator values were transformed to a common range between 0.1 and 1.0 using homothetic transformation equations (1) and (2) [38]. In the context of this study, this is termed as"homothetic transformation method of LSF" (II^L) :

$$Y = 0.1 + \left(\frac{(X - b)}{(a - b)} \right) * 0.9, \qquad (1)$$

$$Z = 1 - \left(\frac{(X - b)}{(a - b)} \right) * 0.9, \qquad (2)$$

Where Y and Z are values of the variables after transformation X is value of the variable to transform and *a* and *b* are the maximum and minimum threshold values of the variable (Table 3). Equation (1) is used for "more is better" scoring function, (2) for "less is better," and a combination of both equations for "optimum is better" scoring function. This LSF approach is therefore described as the "homothetic transformation method" (II^L) .

Table 3: Soil quality indicators, scoring curve, and threshold and baseline limits used for evaluating eight LUSMS in northern Ethiopia

Indicator	Scoring curve	Threshold		Baselinec	Optimumd	Slope at baseline	Source of limits
		Lowera	Upperb				
Sand (%)	Optimum	0	60	L, 30 U, 50	36	0.440	Natural ecosystem, best managed soils
Silt (%)	More is better	0	38	19	—	0.249	Best managed soil, natural ecosystem
Clay (%)	More is better	0	31	15	—	0.268	Natural ecosystem
SAS (%)	More is better	10	60	30	—	0.296	Harris et al. [17]; natural system
BD (g cm−3)	Less is better	1	2.2	1.5	1.2	−0.323	Harris et al. [17]; natural ecosystem
A-horizon depth (cm)	More is better	0	20	10	—	0.318	Natural ecosystem
MWHC (%)	More is better	20	58	30	—	0.198	Gregory et al. [41]; natural system

Parameter							Reference
OCe (%)	More is better	1	6.5	3.5	—	1.046	Kay and Angers [42]; natural ecosystem
CEC (cmolc kg–1)	More is better	6	46	20	—	0.245	Natural ecosystem, best managed soil
TN (%)	More is better	0.05	0.54	0.30	—	25.408	Natural ecosystem,
TP (mg kg–1)	More is better	200	1316	650	—	0.005	Natural ecosystem
Pav (mg kg–1)	More is better	5	29	15	—	0.433	Mausbach and Seybold [43]; natural ecosystem
Zn (mg kg–1)	Optimum	2	20	L, 10 / U, 18	14	0.855	Mausbach and Seybold [43]; natural system
Fe (mg kg–1)	Optimum	10	50	L, 20 / U, 40	26	0.577	Harris et al. [17]; natural system

							Natural ecosystem
Earth worm count per m2	More is better	0	11	5	—	0.470	

[a] Soils at or below the threshold values are prone to structural destabilization, erosion, and low productivity; so the scoring value is 0.

[b] Soils at or beyond this values no further increase in productivity or decrease in erosion rate are achieved the upper threshold; values at and above this level thus receive a score of 1.0.

[c] Values receive a score of 0.5 and are generally regarded as the minimum target values.

[d] The value is given a score of 1.0 if the desired relationship is bell-shaped.

[e] According to Kay and Angers [42], irrespective of soil type if SOC contents are below 1%, it may not be possible to obtain potential yields.

L: lower; U: upper; SAS: soil aggregate stability; BD: bulk density; MWC: maximum water holding capacity; OC: organic carbon; CEC: cation exchangeable capacity; TN: total nitrogen; TP: total phosphorous; Pav: available phosphorous; Zn: available Zinc; Fe: available iron; —: implies not applicable.

The third LSF approach was adopted from Masto et al. [11], Glover et al. [39], and Masto et al. [40]. It is described as:

$$(Y) = \frac{(x - s)}{(t - s)},$$
(3)

$$(Y) = 1 - \frac{(x - s)}{(t - s)},$$
(4)

where Y is the linear score, x is the soil property value, and s and t are lower and upper threshold values (Table3). For values below and above the threshold, the score is zero. Equation (3) was used for "more is better," whereas (4) for "less is better" and a combination of both for "optimum is better." This approach is therefore described as "Glover LSF method" (III^L).

Nonlinear Scoring Functions: Two nonlinear scoring function (NLSF) approaches were also evaluated. The first NLSF approach transformed the indicators using curves constructed with CurveExpert version 1.3 shareware (http://www.flu.org.cn/en/download-79.html) as described by Andrews et al. [13]. The shape of each curve, that is, bell-shaped (midpoint is optimum), sigmoid with an upper asymptote (more is better) or sigmoid with a lower asymptote (less is better), was determined according to agronomic and environmental soil functions using data from undisturbed fields (natural ecosystem), literature values, and knowledge of experts. To develop indicator curve using CurveExpert, it was assumed that levels of activity found in undisturbed soil (natural forest) would have a score at or near 1.0. Polynomial fit model was applied after examining the data by curve finder and then drawing the curve based on the observed values of SQ indicators from undisturbed ecosystem. After interpolated by polynomial Lagrangian interpolation method in the CurveExpert version 1.3, the corresponding transformed value was analyzed for each untransformed indicator value in each LUSMS. The X-axis for the functions represented a site-specific expected range of values of the soil properties. The Y-axis, ranging from 0 to 1, was the transformed score. This approach is termed as the "CurveExpert method of NLSF" (I^N).

The second NLSF used (5) to normalize SQ indicators suggested by Masto et al. [11]; Glover et al. [39]; Masto et al. [40] and is therefore termed as the (II^{N}) "

$$NLSF(Y) = \frac{1}{[1+e^{-b(x-A)}]},\qquad\qquad (5)$$

Where x is the indicator value, A is the baseline or value of the soil property where the score equals 0.5 or about the midpoint between the upper and lower threshold value, and b is the slope. Baseline values are generally regarded as the minimum target value. If soil indicator values are located within the control limits, the system is considered to be in an acceptable state. Conversely, if the values lie outside the threshold limits, the system is considered to be in a state of degradation [11, 40, 44].

Critical values or thresholds were established based on a range of values measured in natural ecosystems, best managed systems, and values adopted from literatures and personal experiences of the researcher to better fit to the local conditions within the catchment (Table 3). This table also presents the indicator scoring curve, baseline, and threshold values used to transform selected SQ indicators. For detailed descriptions of these standard scoring functions the reader is referred to Masto et al. [11]; Glover et al. [39]; Masto et al. [40]; Karlen et al. [45].

Measured values are transformed into unitless scores ranging from 0 to 1 so that scores can be combined and averaged into a single value such that a score of 1 represents the highest potential function for that system; that is, the indicator is nonlimiting with regard to the pertinent soil functions and processes [46]. An advantage of indexing is that important information can often be captured by scoring that might otherwise go undetected when examining only the observed values [13].

Soil Quality Indexing (Step 3)

Two SQ indexing methods are commonly found in literature, for example, Masto et al. [11]; Glover et al. [39]; Masto et al. [40]. The first is unscreened transformation (the additive index) and the second uses principal component analysis (PCA). Both methods were applied to data from each LUSMS. The SQI values were then compared with those from natural forest land systems to assess the degree of soil degradation or improvement. Details of the two SQ indexing methods are summarized below.

(I) Unscreened Transformation Based Soil Quality Indexing (Unscreened-SQI). SQ indicators are integrated into an index (SQI) by summing the scores from individual indicators and dividing by the total number of indicators (i.e., an additive model) as described in Masto et al. [40]

$$nscreend\ SQI = \left(\frac{\sum_{i=1}^{n} S_i}{n} \right) \qquad (6)$$

Where SQI is the soil quality index, S is the linear or nonlinear scored value of individual indicators, and is the number of indicators included in the dataset.

(II) Principal Component Analysis Based Soil Quality Indexing (PCA-SQI). A standard PCA was conducted using all SQ indicators that showed significantly differences among the LUSMS. Under each principal component (PC), only the variables having high factor loadings and eigenvalues >1 that explained at least 5% of the data variations were retained for indexing. Among well-correlated variables within each PC, the variable with the highest correlation coefficient (absolute value) and loading factor was chosen. If the highly weighted variables ($\geq \pm 0.7$ eigenvector) were not well correlated ($r < 0.60$), each was considered important and retained in the PC for SQ indexing. As each PC explains a certain amount of variation within the total dataset, this provides a "weight" for the variables chosen under a given PC. The final PCA based SQI (7) described by Masto et al. [40] was used for this study:

$$PCA-SQI = \sum_{i=1}^{n} W_i S_i,$$ (7)

where PCA-SQI is principal component analysis (PCA) based soil quality index, is the PCA weighing factor equal to the ratio of variance of each factor to total cumulative variance coefficients in the equation, and is scored value of each SQ indicator.

Evaluation of Soil Quality Indexing Methods

The SQ indexing methods were evaluated using sensitivity analysis described by Masto et al. [11] as

$$Sensitivity(S) = \frac{SQL_{(max)}}{SQL_{(min)}},$$ (8)

where $SQL_{(max)}$ and $SQL_{(min)}$ are the maximum and minimum SQI observed under each scoring procedure using each dataset selection methods. The SQ indexing method with higher value of sensitivity is more preferable as this is sensitive to perturbations and management practices [21].

Data Analysis

Data were analyzed using statistical software package of SPSS 18.0 [47]. One-way analysis of variance (ANOVA) was performed to determine the effects of LUSMS on SQI. Mean SQI was tested for its level of significance at probability level $(P) \leq 0.05$. Data were also analyzed using correlation and factor analysis. A PCA was used to examine the relationship among the 24 SQ indicators by statistically grouping them into four PC factors through the varimax rotation procedure. Varimax rotation with Kaiser Normalization was used because this results in a factor pattern that highly loads into one factor [48]. Communalities estimate the portion of variance of each soil attribute that explains for the factors. A high communality for a soil attribute indicates a high proportion of the variance explained in the factor. Less importance should be ascribed to soil attributes with low communalities when interpreting the factors [48].

RESULTS AND DISCUSSION

Grouping Soil Quality Indicators

A moderate to strong correlation $(r > 0.7)$ among many SQ indicators within the different LUSMS was observed, indicating a multicollinearity effect (data not shown). Factor analysis can help reduce the number of indicators analyzed needed for indexing by identifying components that best account for the variability and thus minimizing data redundancy (multicollinearity effect). The 25 SQ indicators analyzed to evaluate the eight LUSMS were grouped using a PCA. This resulted in four principal component (PC) groups that best explained variability in the data (Table 4). Communalities of the SQ indicators (Table 4) show that individual indicators accounted for 64 to 97% of the variance. Indicators with high communality get preference over those with low communality [49]. When combined, the first four PCs factors with eigenvalues >1 explained about 88% of the soil variability among the eight LUSMS. The first two PCs accounted for about 52% of the variance, indicating that these would be potential components to assess SQ effects within the LUSMS.

Table 4: Principal component analysis results using 25 potential soil quality indicators to evaluate eight LUSMS in the Mai-Negus catchment, northern Ethiopia

Eigenvector	Principal component, PC				Communalities
	1	2	3	4	
Organic carbon, OC	0.86	0.29	0.12	0.28	0.92
Total nitrogen, TN	0.86	0.21	0.04	0.36	0.92
Earthworm per m2, EW	0.79	0.27	0.28	0.23	0.82
Porosity	0.74	0.38	0.40	0.30	0.94
Zinc, Zn	0.70	0.46	0.31	0.16	0.82
Water-holding capacity, WHC	0.69	0.47	0.47	0.20	0.96

Available phosphorus, Pav	0.69	0.34	0.31	0.11	0.71
A-horizon depth, AHD	0.64	0.37	0.48	0.41	0.94
Soil aggregate stability, SAS	0.60	0.45	0.52	0.33	0.94
Exchangeable potassium, K	0.27	0.83	0.28	0.17	0.86
Cation exchangeable capacity, CEC	0.42	0.76	0.51	0.29	0.95
Exchangeable calcium, Ca	0.41	0.72	0.40	0.28	0.93
Total phosphorous, TP	0.45	0.70	0.38	0.11	0.85
Exchangeable magnesium, Mg	0.53	0.68	0.38	0.24	0.94
Sum base forming cations, SBF	0.45	0.63	0.40	0.26	0.97
Silt	0.19	0.39	0.84	0.04	0.90
Dry bulk density, DBD	−0.40	−0.38	−0.78	−0.30	0.94
pH	0.22	0.26	0.74	0.26	0.82
Sand	−0.34	−0.36	−0.67	−0.30	0.91
EC	0.38	0.44	0.55	−0.03	0.64
Exchangeable sodium, Na	−0.18	−0.18	0.01	−0.57	0.82
Iron, Fe	−0.30	−0.08	−0.13	0.86	0.85
Exchangeable sodium percentage, ESP	−0.17	−0.25	−0.49	−0.59	0.97
Clay	0.51	0.13	0.21	0.63	0.72
Eigenvalue	7.26	5.16	4.79	3.82	—
Variance (%)	30.26	21.51	19.96	15.90	—
Cumulative variance (%)	30.26	51.77	71.73	87.63	—

Boldface eigenvector values correspond to the PCs highly weighted variables examined for the index.

Bold-italic factors correspond to the indicators retained in the SQ index. The weight of the variables included in the index was decided using the variance of each factor.

PCA-SQI = 0.303OC + 0.0303TN + 0.215CEC + 0.215TP + 0.200silt + 0.200DBD + 0.159Fe.

Normalized   PCA-SQI   =

(0.303OC + 0.0303TN + 0.215CEC + 0.215TP + 0.200DBD + 0.159Fe) / (1.595) = 0.190OC + 0.190TN + 0.135CEC + 0.135TP + 0.125silt + 0.125DBD + 0.100Fe.

Eigenvectors for the first four PCs (Table 4) show that OC, TN, EW, porosity, and Zn were retained for PC1. The correlation coefficient between TN and the other variables was less than 0.6, so TN was retained in PC1. However, the high correlation coefficient ($r > 0.87$) of OC with the other high loading variables suggested that OC was most important so it too was retained in PC1 for SQ indexing. Several literature references show that EW populations are highly influenced by organic matter availability [50, 51]. Furthermore, according to Jenkinson [52], soil microbial biomass comprises 1 to 4% of the total OC and 2 to 6% of the total organic nitrogen. Based on this information, we chose to exclude EW data from PC1 since its contribution is explained by soil organic matter. For such reasons, PC1 is referred to as the "soil organic matter factor."

Soil CEC, TP, exchangeable K and Ca were the highly loaded factors attributed to PC2. The CEC values were strongly correlated ($r = 0.85$) with K, and Ca so CEC was selected as the preferred variable for retention in PC2. The correlation coefficient of TP with the other variables was <0.6 and since the cutoff is $r = 0.7$, TP was also retained in PC2. Phosphorus is frequently a limiting factor for crop production in northern Ethiopia soils, so its inclusion in a SQ index is logical for assessing SQ degradation within the various LUSMS. Based on these two factors, PC2 is referred to as the "soil macro-nutrient factor."

The only highly loaded variables in PC3 were silt, DBD, and pH. The correlation of silt with the highly loaded variables was less than 0.6, so it was retained in PC3 for index development. A similar analysis showed a higher partial correlation for DBD ($r > 0.80$), indicating that it was also a preferred indicator for retention in PC3. Both silt and DBD also showed higher communality effect when compared to pH. These indicators also influenced the SQ in an opposite direction (Table 4); their inclusion in any SQ indexing is crucial to assess variability associated with the various LUSMS. Again, based on the critical indicators, PC3 is referred to as the "soil physical property factor." The fourth PC is referred to as the "soil micro-nutrient factor," because Fe is the highly loaded variable (Table 4).

Based on the four PCs, a composite PCA-SQI consisting of soil OC, TN, CEC, TP, silt, DBD, and Fe was chosen to assess SQ variability among the LUSMS. Weighting factors were developed based on the percent variation explained by the first four PCs (Table 4), resulting in a final normalized PCA based SQI equation:

$$PCA - SQL = 0.190_{OC} + 0.190_{IN} + 0.135_{CBC} + 0.135_{IP}$$
$$+0.125_{silt} + 0.125_{DBD} + 0.100_{Fe},$$

(9)

Where PCA-SQI is a PCA based soil quality index and is the score (linear or nonlinear) for each variable (Table4), with coefficients based on the variance accounted for by each PC.

Integration of Soil Quality Indicators into a Soil Quality Index

Unscreened Transformation Based Soil Quality Index (Unscreened-SQI)

An unscreened-SQI using the PCA and EO selected minimum datasets showed the highest SQI values for the II^L (homothetic transformation method) and I^N (CurveExpert method of NLSF) models used to compare the eight LUSMS (Tables 5 and 6). For all comparisons, the nonlinear unscreened-SQI showed higher values than the linear scored functions. In addition, unscreened-SQI values for both LSF and NLSF methods were greater for EO minimum datasets than for PCA selected datasets.

Table 5: Soil quality indices developed using a PCA selected dataset to compare eight LUSMS in the Mai-Negus catchment of northern Ethiopia

LUSMS	Unscreened SQI					PCA-SQI					Meanb
	LSFa			NLSFa		LSF			NLSF		
	I^L	II^L	III^L	I^N	II^N	I^L	II^L	III^L	I^N	II^N	
LS1	0.913a	0.941a	0.899a	0.956a	0.953a	0.913a	0.925a	0.881a	0.973a	0.964a	0.932a
LS2	0.823a	0.843a	0.793b	0.920a	0.916a	0.826b	0.845a	0.795b	0.928a	0.921a	0.861a
LS3	0.621bc	0.659b	0.626c	0.747bcd	0.741bcd	0.619c	0.658b	0.593c	0.773bc	0.754bc	0.679b
LS4	0.648b	0.673b	0.627c	0.780bc	0.777bc	0.618c	0.608bc	0.587c	0.774bc	0.772bc	0.686b
LS5	0.539cd	0.582bc	0.522d	0.689cd	0.686cd	0.484d	0.535cd	0.470d	0.696cd	0.692cd	0.589c
LS6	0.506d	0.532cd	0.510d	0.655de	0.652de	0.463d	0.514d	0.453d	0.656d	0.653d	0.559c
LS7	0.639b	0.676b	0.660c	0.787b	0.784b	0.597c	0.638b	0.591c	0.791b	0.781b	0.694b
LS8	0.410e	0.475d	0.405e	0.562e	0.559e	0.347e	0.418e	0.341e	0.513e	0.546e	0.458d
mean	0.637	0.672	0.630	0.762	0.759	0.609	0.642	0.589	0.763	0.760	0.682
LSD ($P = 0.05$)	0.095	0.099	0.091	0.097	0.096	0.086	0.082	0.076	0.089	0.089	0.083

Means followed by different letters in the same column are significantly different at $P = 0.05$; LSD: least significance difference.

Details of LSF and NLSF descriptions can be found in Section 2.4 (Step2).

This shows overall mean of the different scoring and indexing methods across the columns for the same LUSMS as tested statistically.

LSF: linear scoring function; NLSF: nonlinear scoring function; I^L: Liebig method LSF; II^L: Homothetic transformation method of LSF; III^L: Glover method LSF; I^N: CurveExpert method NLSF; II^N: Glover method NLSF; LUSMS: land use and soil management systems; SQI: soil quality index; PCA: principal component analysis; LS1: natural forest (reference); LS2: plantation of protected area; LS3: pasture land system; LS4: teff-faba bean rotation; LS5: teff-barley/wheat rotation; LS6: teff-monocropping; LS7: Maize monocropping; LS8: uncultivated-marginal land soil system.

Table 6: Soil quality indices developed using an EO selected dataset to compare eight LUSMS in the Mai-Negus catchment of northern Ethiopia

LUSMS	Unscreened SQI					PCA-SQI					
	LSFa			NLSFa		LSF			NLSF		Meanb
	IL	IIL	IIIL	IN	IIN	IL	IIL	IIIL	IN	IIN	
LS1	0.937a	0.943a	0.925a	0.966a	0.964a	0.796a	0.799a	0.779a	0.818a	0.816a	0.874a
LS2	0.828b	0.831b	0.786b	0.920a	0.918a	0.724b	0.723b	0.690b	0.783a	0.781a	0.799b
LS3	0.707c	0.720c	0.673c	0.830b	0.828b	0.612c	0.622c	0.584c	0.708b	0.666b	0.695c
LS4	0.652c	0.662c	0.610c	0.775b	0.773b	0.556d	0.565d	0.522d	0.651c	0.650b	0.642d
LS5	0.529d	0.552d	0.491d	0.658c	0.656c	0.448e	0.468e	0.417e	0.548d	0.546c	0.531e
LS6	0.501d	0.526d	0.461d	0.621c	0.619c	0.428e	0.450e	0.397e	0.497e	0.517c	0.502e
LS7	0.657c	0.670c	0.622c	0.773b	0.771c	0.545d	0.558d	0.514d	0.653c	0.651b	0.641d
LS8	0.397e	0.434e	0.361e	0.511d	0.509d	0.335f	0.369f	0.307f	0.420f	0.418d	0.406f
mean	0.651	0.667	0.616	0.756	0.754	0.556	0.569	0.526	0.635	0.631	0.636
LSD ($P = 0.05$)	0.066	0.059	0.067	0.060	0.060	0.044	0.038	0.042	0.042	0.051	0.048

Means followed by different letters in the same rows are significantly different at ; LSD: least significance difference.

Details of LSF and NLSF description can be found in Section 2.4 (Step2).

This shows the overall mean of the different scoring and indexing methods across the columns for the same LUSMS tested statistically.

Explanations of abbreviations are similar to footnotes under Table 5.

Among LUSMS, LS1 had a significantly higher ($P = 0.05$) unscreened-SQI value whereas a lower value in LS8 using either the PCA or EO selected datasets. The general SQI pattern among the LUSMS was LS1 > LS2 > LS7 > LS4 > LS3 > LS5 > LS6 > LS8, (Tables 5 and 6). Mean unscreened-SQI differences between the LS1 reference soils and average unscreened-SQI values for PCA and EO datasets showed the following soil degradation levels: LS2 (−8%), LS3 (−20%), LS7 (−24%), LS4 (−26%), LS6 (−38%), LS5 (−45%), and LS8 (−58%). Generally, this study demonstrated that LS2 followed by LS3 and LS7 is more advantageous in maintaining SQ than the other LUSMS in the catchment.

PCA Based Soil Quality Indices (PCA-SQI)

For both PCA and EO databases, NLSF methods resulted in higher SQI values than LSF (Tables 5 and 6). In contrast to unscreened-SQI values, PCA-SQI values derived from PCA selected datasets were higher than those derived using EO datasets. Once again,the II^L and I^N indexing methods resulted in the highest PCA-SQI values for both data selection methods (Tables 5 and 6). Among the LUSMS, the PCA-SQI value for LS1 was significantly higher () than the other land uses although it has been just slightly better than LS2 and LS3. LS8 had the lowest PCA-SQI value followed closely by LS5 and LS6 (Tables 5 and 6). Soil degradation based on PCA-SQI differences between each LUSMS and the LS1 reference averaged −6, −21, −23, −24, −34, −36, and −59% for LS2, LS3, LS7, LS4, LS5, LS6, and LS8, respectively, when computed using average values for PCA and EO derived datasets.

Dataset Selection, Scoring, and Indexing Methods Comparison

Using NLSF, PCA-SQI values were greater than unscreened-SQI values but with LSF methods unscreened-SQI values were higher. Comparing MDS selection methods, unscreened-SQI values derived using an EO dataset were higher for both LSF and NLSF methods (Tables 5 and 6).The II^L and I^N scoring methods resulted in higher SQI values than the other methods. Overall, for almost all LUSMS, NLSF values were generally greater than LSF values. Mean SQI values using PCA selected datasets ranged from 0.458 (LS8) to 0.932 (LS1) (Table 5). The overall average using EO selected datasets ranged from 0.406 (LS8) to 0.874 (LS1) (Table 6). This implies that mean SQI values calculated from both indexing methods using EO datasets resulted in lower index values than with PCA selected datasets. Therefore, this study concludes that adopting the PCA dataset selection method is the best for evaluating the soil degradation status of LUSMS in northern Ethiopia and under similar environmental conditions. The use of a PCA dataset selection method also minimizes any disciplinary bias that may be associated with the EO selection method.

Evaluation of Indexing Methods

The unscreened-SQI values varied from 0.410 to 0.933 and 0.397 to 0.954 for the PCA and EO dataset selection methods, respectively. Likewise, PCA-SQI values ranged from 0.347 to 0.934 and 0.335 to 0.801 for the two dataset selection methods, respectively (Table 7). A sensitivity analysis showed that both indexing and dataset selection methods influenced the SQI values for the various LUSMS throughout the catchment (Table 7). It also confirmed that indexing, scoring, and MDS selection can be used to evaluate soil degradation among LUSMS in the Mai-Negus catchment of northern Ethiopia. However, the sensitivity analyses do not agree with findings reported by Masto et al. [11] and Andrews et al. [53] who favored NLSF over LSF. One reason for this difference is that the previous reports were based on single scoring functions from LSF and NLSF methods and their sensitivity was not evaluated. Also, many of the LSF and NLSF approaches evaluated in this project were not addressed at the same time and place in the previous studies.

Table 7: Dataset selection, scoring function, and indexing method comparisons for eight LUSMS in the Mai-Negus catchment of northern Ethiopia

SQ indexing method	Linear scoring function						Nonlinear scoring function			
	IL		IIL		IIIL		IN		IIN	
	Range	S	Range	S	Range	S	Range	S	Range	S
Unscreened-SQIa	0.410–0.913	2.227	0.475–0.941	1.981	0.405–0.899	2.220	0.562–0.956	1.701	0.559–0.953	1.705
PCA-SQIa	0.347–0.913	2.631	0.418–0.923	2.208	0.341–0.881	2.584	0.513–0.973	1.898	0.546–0.964	1.766
Unscreened-SQIb	0.397–0.937	2.360	0.434–0.943	2.173	0.361–0.925	2.562	0.511–0.966	1.890	0.508–0.964	1.897
PCA-SQIb	0.335–0.796	2.376	0.369–0.799	2.165	0.307–0.779	2.537	0.420–0.818	1.952	0.418–0.816	1.948
Mean	—	2.399	—	2.132	—	2.476	—	1.859	—	1.830

[a] denotes minimum dataset chosen using principal component analyses (PCA).

[b] denotes minimum dataset chosen by expert opinion (EO).

IL: Liebig method LSF; IIL: Homothetic transformation method of LSF; IIIL: Glover method LSF; IN: CurveExpert method NLSF; IIN: Glover method NLSF; S: Sensitivity analysis; LSF: linear scoring function; NLSF: nonlinear scoring function.

Synthesis of Dataset Selection, Scoring, and SQ Indexing Methods

Dataset selection methods (PCA versus EO), scoring functions, and indexing approaches all influenced the SQI values used to compare LUSMS in northern Ethiopia (Tables 5, 6 and 7). This study results were consistent with previous studies that also showed these factors contributed to variability in SQI values [11, 16, 34, 35]. Arguments favoring EO method for selection of a MDS stress its focus on sustainable management goals [11, 36,45], but many researchers have also relied on statistical techniques such as PCA (e.g., [11, 16, 53]). This study confirmed PCA selection could reduce expert biases as compared to an EO method. Andrews et al. [16] reported that EO and PCA dataset selection methods produced almost similar SQI values, but PCA-SQI values using PCA and NLSF methods in this study were higher. This was not true for unscreened-SQI which was obtained by summing scores for individual indicators.

Many studies have shown that NLSF resulted in higher SQI values than LSF [11, 16], but a sensitivity analysis showed that the I^L method explained variation better than the NLSF method for both indexing methods. This was followed by another LSF method denoted as III^L. Several strategies for integrating scored indicators values into an overall SQI have been proposed, but no approach has received universally acceptance [11].

SQI values were segregated into physical, chemical, and biological components (Figure 3) to show the component effects on SQ and thus soil degradation within each LUSMS in the catchment. The individual factors also permit comparisons among the scoring methods that help illustrate differences among them. Finally, although such comparisons are interesting, they are not adequate for assessing SQ effects of LUSMS because numerous interactions also contribute to the overall SQI values [13, 15]. This is also a major reason that use of a single SQ indicator regardless of how its scored or interpreted is not suitable for comparing overall soil degradation dynamics among LUSMS.

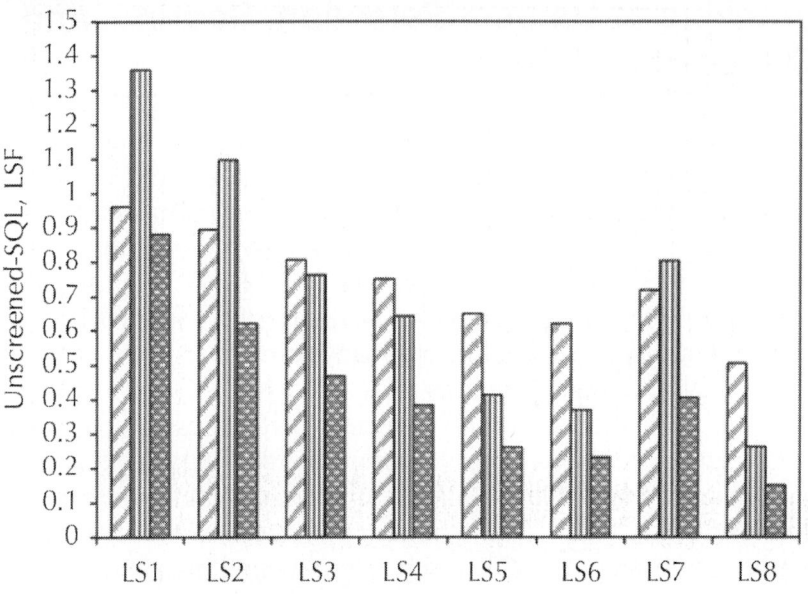

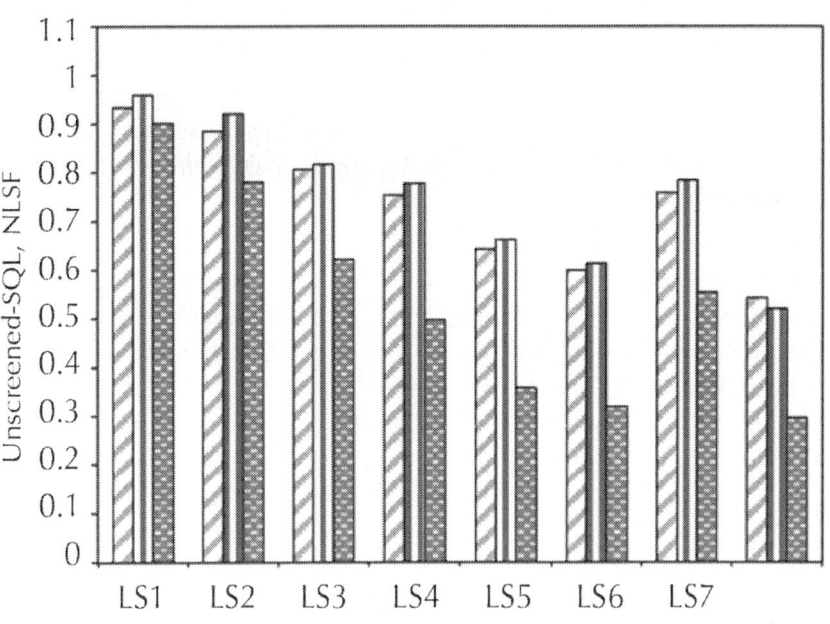

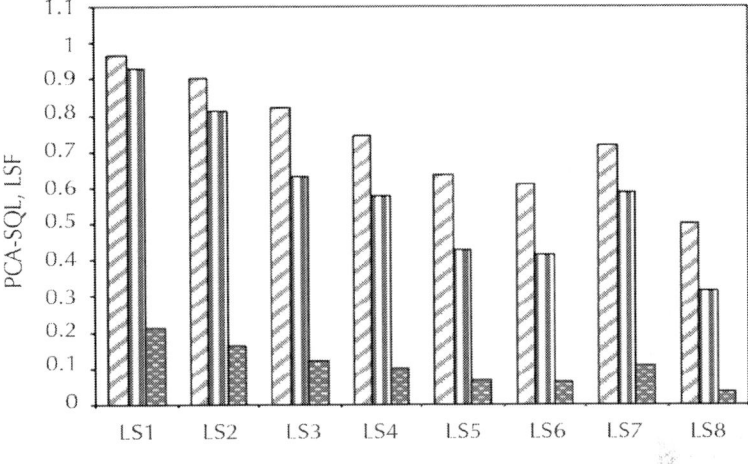

☐ Physical index
▥ Chemical index
▨ Biological index

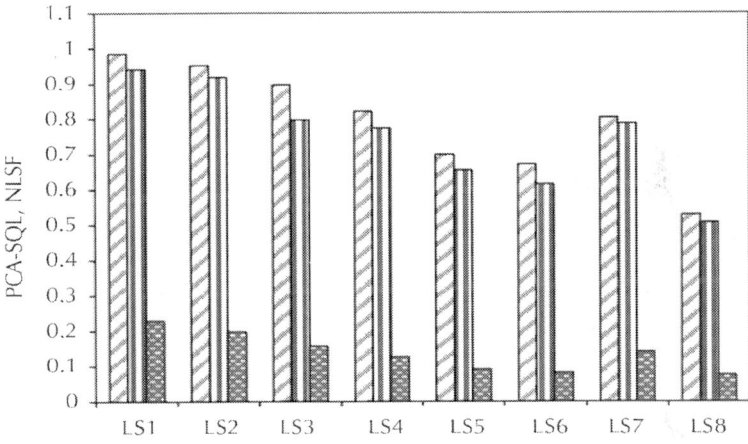

☐ Physical index
▥ Chemical index
▨ Biological index

Figure 3: LUSMS effects on mean physical, chemical, and biological soil quality indicator scores for different indexing methods within the Mai-Negus catchment of northern Ethiopia.

Several studies have shown that agroforestry can be a viable option for restoration of degraded land as it improves soil indicators such as organic matter, nitrogen cycling, soil structure, and biological activity [54–56]. Others have added that extensive rooting systems and protective canopies of the Leucaena leeucocephala andSesbania sesban trees can protect the soil from erosion and create favorable conditions for plant and microbial growth [56, 57]. Litter-fall and fine roots decomposition both contribute to these processes. This study shows higher SQI value for LS1 and LS2 because of better vegetation cover (natural and plantation), which is consistent with previous reports [54–57]. Overall, this study confirmed that SQI values can be developed and used as tools for early identification of land degradation particularly the soil component. They can also be used to support decision making processes with regard to sustainable soil management planning in the context of different LUSMS at catchment scale.

CONCLUSIONS

This study demonstrated that natural forest land systems (LS1) have relatively good soil quality (SQ), whereas uncultivated marginal land systems (LS8) have a seriously degraded soil within the Mai-Negus catchment of northern Ethiopia. Areas being managed in a teff-barley/wheat rotation land system (LS5) and teff monocropping (LS6) also had not reasonably good SQ. SQI values for LS8 and LS6 provided an early warning regarding the severity of soil degradation since more than 50% of the original soil is degraded when compared to LS1 (reference soil). The PCA-SQI method resulted in a higher SQI value than the unscreened-SQI for NLSF when datasets were selected using a PCA method. However, the reverse was true for the datasets selected using an EO method. The NLSF methods resulted in higher SQI values than LSF methods for all LUSMS regardless of the SQ indexing method. Among NLSF, the Curve Expert method (I^N) showed the highest SQI for the dataset selected by the PCA method. However, a sensitivity analysis indicated that PCA-SQI with the LSF (denoted as I^L or the "Liebig method") and a PCA derived dataset had the highest values. The lowest sensitivity values were associated with an unscreened-SQI using NLSF method denoted as II^N (Glover method NLSF) using PCA derived datasets. This study also indicated that the PCA-SQI method with a PCA selected dataset and a LSF designated as I^L was the most sensitive for assessing differences among the eight LUSMS in northern Ethiopia. Based on this study result, the researcher concludes that use of relevant datasets, scoring functions, and indexing methods are all important factors influencing SQI values within the eight LUSMS. Also, for improving degraded soils and maintaining SQ, appropriate interventions within the various LUSMS should be identified, prioritized, and monitored using SQI values as decision aids.

ACKNOWLEDGMENTS

The author gratefully acknowledges the financial support by DAAD/ GIZ (Germany) through the Center for Development Research (ZEF), University of Bonn (Germany), and field work supported by Aksum University (Ethiopia). The author also highly appreciates the cooperation of the participant farmers and assistance offered by the local administration and extension agents during the field work. The author is also grateful to the anonymous reviewers for their comments which helped in improving this paper.

REFERENCES

1. H. Eswaran, R. Lal, and P. F. Reich, "Land degradation: an overview," in Response to Land Degradation, E. M. Bridges, I. D. Hannam, L. R. Oldeman, F. W. T. Penning De Vries, J. S. Scherr, and S. Sombatpanit, Eds., pp. 20–35, Science Publishers, Enfield, NH, USA, 2001.

2. G. Girmay, B. R. Singh, H. Mitiku, T. Borresen, and R. Lal, "Carbon stocks in Ethiopian soils in relation to land use and soil management," Land Degradation and Development, vol. 19, no. 4, pp. 351–367, 2008.

3. R. Lal, "Soil erosion problems on alfisols in Western Nigeria, VI. Effects of erosion on experimental plots,"Geoderma, vol. 25, no. 3-4, pp. 215–230, 1981.

4. R. Lal, T. M. Sobecki, T. Iivari, and J. M. Kimble, Soil Degradation in the United States: Extent, Severity and Trends, CRC Press, Boca Raton, Fla, USA, 2003.

5. M. A. Denboba, Forest conversion—soil degradation—farmers' perception nexus: implications for sustainable land use in the Southwest of Ethiopia [Ph.D. thesis], University of Bonn, Bonn, Germany, 2005.

6. J. W. Doran, "Soil health and global sustainability: translating science into practice," Agriculture, Ecosystems and Environment, vol. 88, no. 2, pp. 119–127, 2002.

7. H. Hurni, K. Tato, and G. Zeleke, "The implications of changes in population, land use, and land management for surface runoff in the Upper Nile Basin Area of Ethiopia," Mountain Research and Development, vol. 25, no. 2, pp. 147–154, 2005.

8. A. Glanz, Saving Our Soil: Solutions for Sustaining Earth's Vital Resource, Johnson Books, Boulder, Colo, USA, 1995.

9. J. F. Parr, R. I. Papendick, S. B. Hornick, and R. E. Meyer, "Soil quality: attributes and relationship to alternative and sustainable agriculture," American Journal of Alternative Agriculture, vol. 7, no. 1-2, pp. 5–11, 1992.

10. M. R. Carter, "Soil quality for sustainable land management: organic matter and aggregation interactions that maintain soil functions," Agronomy Journal, vol. 94, no. 1, pp. 38–47, 2002

11. R. E. Masto, P. K. Chhonkar, D. Singh, and A. K. Patra, "Alternative soil quality indices for evaluating the effect of intensive cropping, fertilisation and manuring for 31 years in the semi-arid soils of India,"Environmental Monitoring and Assessment, vol. 136, no. 1–3, pp. 419–435, 2008.

12. Z. Sakbaeva, V. Acosta-Martinez, J. Moore-Kucera, W. Hudnall, and K. Nuridin, "Interactions of soil order and land use management on soil properties in the Kukart watershed, Kyrgyzstan," Applied and Environmental Soil Science, vol. 2012, Article ID 130941, 11 pages, 2012.

13. S. S. Andrews, D. L. Karlen, and C. A. Cambardella, "The soil management assessment framework: a quantitative soil quality evaluation method," Soil Science Society of America Journal, vol. 68, no. 6, pp. 1945–1962, 2004.

14. W. J. Elliot, D. P. Dumroese, and P. R. Robichaud, "The effects of forest management on erosion and soil productivity," in Soil Quality and Soil Erosion, R. Lal, Ed., pp. 195–209, CRC Press, Boca Raton, Fla, USA, 1999.

15. M. C. Amacher, K. P. O'Neill, and C. H. Perry, "Soil vital signs: a new soil quality index (SQI) for assessing forest soil health," Research Paper of USDA, Forest Service, Rocky Mountain Research Station, 2007.

16. S. S. Andrews, D. L. Karlen, and J. P. Mitchell, "A comparison of soil quality indexing methods for vegetable production systems in Northern California," Agriculture, Ecosystems and Environment, vol. 90, no. 1, pp. 25–45, 2002.

17. R. F. Harris, D. L. Karlen, and D. J. Mulla, "A conceptual framework for assessment and management of soil quality and health," in Methods for Assessing Soil Quality, J. W. Doran and A. J. Jones, Eds., vol. 49, pp. 61–82, Soil Science Society of America, Madison, Wis, USA, 1996.

18. S. S. Andrews and C. R. Carroll, "Designing a soil quality assessment tool for sustainable agroecosystem management," Ecological Applications, vol. 11, no. 6, pp. 1573–1585, 2001.

19. B. K. Gugino, O. J. Idowu, R. R. Schindelbeck et al., Cornell Soil Health Assessment Training Manual, Cornell University, Geneva, NY, USA, 2nd edition, 2009.

20. Food and Agriculture Organization of the United Nations, The Soil and Terrain Database for Northeastern Africa (CDROM), FAO, Rome, Italy, 1998.

21. J. A. Burger and D. L. Kelting, "Soil quality monitoring for assessing sustainable forest management," in The Contribution of Soil Science to the Development and Implementation of Criteria and Indicators of Sustainable Forest Management, J. M. Gigham, Ed., vol. 53, pp. 17–45, Soil Science Society of America, Madison, Wis, USA, 1998.

22. G. W. Gee and J. W. Bauder, "Particle-size analysis," in Methods of Soil Analysis, A. Klute, Ed., Part 1, pp. 383–411, America Society of Agronomy, Soil Science Society of America, Madison, Wis, USA, 1986.

23. G. R. Blake and K. H. Hartge, "Bulk density," in Methods of Soil Analysis, A. Klute, Ed., Agronomy Monograph 9, Part 1, pp. 363–375, America Society of Agronomy, Madison, Wis, USA, 1986.

24. The Nature and Properties of Soils, Prentice-Hall, Upper Saddle River, NJ, USA, 13th edition, 2002, edited by N.C. Brady and R.R. Weil.

25. M. H. Beare and R. Russell Bruce, "A comparison of methods for measuring water-stable aggregates: implications for determining environmental effects on soil structure," Geoderma, vol. 56, pp. 87–104, 1993.

26. T. C. Baruah and H. P. Barthakur, A Text Book of Soil Analysis, Vikas Publishing House, New Delhi, India, 1999.

27. G. W. Thomas, "Soil pH and soil acidity," in Methods of Soil Analysis: Chemical Methods, D. L. Sparks, Ed., part 3, pp. 475–490, Soil Science Society of America, Madison, Wis, USA, 1996.

28. J. M. Bremmer and C. S. Mulvaney, "Nitrogen total," in Method of Soil Analysis, Part 2. Chemical and Microbiological Properties, A. L. Page, Ed., Agronomy Monograph 9, pp. 595–624, America Society of Agronomy, Madison, Wis, USA, 1982.

29. S. R. Olsen and L. E. Sommers, "Phosphorus," in Method of Soil Analysis: Chemical and Microbiological Properties, A. L. Page, Ed., Agronomy Monograph 9, part 2, pp. 403–430, America Society of Agronomy, Madison, Wis, USA, 1982.

30. J. M. Anderson and J. S. I. Ingram, Tropical Soil Biology and Fertility, A Handbook of Methods, CAB International, Wallingford, UK, 1993.

31. J. D. Rhoades, "Cation exchange capacity," in Methods of Soil Analysis, A. L. Page, R. H. Miller, and D. R. Keeney, Eds., Agronomy Monograph 9, part 2, pp. 149–157, America Society of Agronomy, Madison, Wis, USA, 1982.

32. M. S. Coyne and J. A. Thompson, Math for Soil Scientists, Thomson Delmar Learning, Clifton Park, NY, USA, 2006.

33. S. S. Andrews, Sustainable agriculture alternatives: ecological and managerial implications of poultry litter management alternatives applied for agronomic soils [Ph.D. dissertation], University of Georgia, Athens, Ga, USA, 1998.

34. J. W. Doran and T. B. Parkin, "Defining and assessing soil quality," in Defining Soil Quality for a Sustainable Environment, J. W. Doran, D. G. Coleman, D. F. Bezddick, and B. A. Stewart, Eds., pp. 3–22, Soil Science Society of America, Madison, Wis, USA, 1994.

35. J. W. Doran and T. B. Parkin, "Quantitative indicators of soil quality: a minimum data set," in Methods for Assessing Soil

Quality, J. W. Doran and J. Jones, Eds., pp. 25–38, Soil Science Society of America, Madison, Wis, USA, 1996.

36. W. E. Larson and F. J. Pierce, "The dynamics of soil quality as a measure of sustainable management," inDefining Soil Quality for Sustainable Environment, J. W. Doran, D. G. Coleman, D. F. Bezddick, and B. A. Stewart, Eds., pp. 37–52, Soil Science Society of America, Madison, Wis, USA, 1994.

37. M. A. Liebig, G. Varvel, and J. Doran, "A simple performance-based index for assessing multiple agroecosystem functions," Agronomy Journal, vol. 93, no. 2, pp. 313–318, 2001.

38. E. Velasquez, P. Lavelle, and M. Andrade, "GISQ, a multifunctional indicator of soil quality," Soil Biology and Biochemistry, vol. 39, no. 12, pp. 3066–3080, 2007.

39. J. D. Glover, J. P. Reganold, and P. K. Andrews, "Systematic method for rating soil quality of conventional, organic, and integrated apple orchards in Washington State," Agriculture, Ecosystems and Environment, vol. 80, no. 1-2, pp. 29–45, 2000.

40. R. E. Masto, P. K. Chhonkar, D. Singh, and A. K. Patra, "Soil quality response to long-term nutrient and crop management on a semi-arid Inceptisol," Agriculture, Ecosystems and Environment, vol. 118, pp. 130–142, 2007.

41. P. J. Gregory, L. P. Simmonds, and C. J. Pilbeam, "Soil type, climatic regime, and the response of water use efficiency to crop management," Agronomy Journal, vol. 92, no. 5, pp. 814–820, 2000.

42. B. D. Kay and D. A. Angers, "Soil structure," in Handbook of Soil Science, M. E. Summer, Ed., pp. 229–269, CRC Press, New York, NY, USA, 1999.

43. M. J. Mausbach and C. A. Seybold, "Assessment of soil quality," in Soil Quality and Agricultural Sustainability, R. Lal, Ed., pp. 33–43, Sleeping Bear Press, Chelsea, Mich, USA, 1998.

44. D. L. Karlen and D. E. Scott, "A framework for evaluating physical and chemical indicators of soil quality," in Defining Soil Quality for a Sustainable Environment, J. W. Doran, D. C. Coleman, D. F. Bezdicek, and B. A. Stewart, Eds., pp. 53–72, Soil Science Society of America, Madison, Wis, USA, 1994.

45.   D. L. Karlen, N. C. Wollenhaupt, D. C. Erbach et al., "Crop residue effects on soil quality following 10-years of no-till corn," Soil and Tillage Research, vol. 31, no. 2-3, pp. 149–167, 1994.

46.   C. A. Seybold, M. J. Mausbach, D. L. Karlen, and H. H. Rogers, "Quantification of soil quality," in Soil Processes and the Carbon Cycle, R. Lal, J. M. Kimble, R. F. Follett, and B. A. Stewart, Eds., pp. 387–404, CRC Press, Washington, DC, USA, 1997.

47.   Statistical Package for Social Sciences, "Release 18.0.," SPSS, 2011.

48.   J. J. Brejda, D. L. Karlen, J. L. Smith, and D. L. Allan, "Identification of regional soil quality factors and indicators: II. Northern Mississippi Loess Hills and Palouse Prairie," Soil Science Society of America Journal, vol. 64, no. 6, pp. 2125–2135, 2000

49.   R. A. Johnson and D. W. Wichern, Applied Multivariate Statistical Analysis, Prentice-Hall, Englewood Cliffs, NJ, USA, 1992.

50.   J. P. E. Anderson and K. H. Demsch, "Quantities of plant nutrients in the microbial biomass of selected soils," Soil Science, vol. 130, pp. 211–216, 1980.

51.   G. P. Sparling, "Ratio of microbial biomass carbon to soil organic carbon as a sensitive indicator of changes in soil organic matter," Australian Journal of Soil Research, vol. 30, no. 2, pp. 195–207, 1992.

52.   D. S. Jenkinson, "Determination of microbial biomass carbon and nitrogen in soil," in Advances in Nitrogen Cycling in Agricultural Ecosystems, J. R. Wilson, Ed., pp. 368–386, CAB International, Wallingford, UK, 1988.

53.   S. S. Andrews, J. P. Mitchell, R. Mancinelli et al., "On-farm assessment of soil quality in California's Central Valley," Agronomy Journal, vol. 94, no. 1, pp. 12–23, 2002.

54.   R. F. Fisher, "Soil organic matter: clue or conundrum," in Carbon Forms and Functions in Forest Soils, W. H. McFee and J. M. Kelly, Eds., pp. 1–11, Soil Science Society of America, Madison, Wis, USA, 1995.

55.   B. Kaur, S. R. Gupta, and G. Singh, "Soil carbon, microbial activity and nitrogen availability in agroforestry systems on moderately alkaline soils in Northern India," Applied Soil Ecology, vol. 15, no. 3, pp. 283–294, 2000.

56. Z. Filip, "International approach to assessing soil quality by ecologically-related biological parameters,"Agriculture, Ecosystems and Environment, vol. 88, no. 2, pp. 169–174, 2002.

57. T. Yan, L. Yang, and C. D. Campbell, "Microbial biomass and metabolic quotient of soils under different land use in the Three Gorges Reservoir area," Geoderma, vol. 115, no. 1-2, pp. 129–138, 2003.

Chapter 9

Effect of Tillage Practices on Soil Properties and Crop Productivity in Wheat-Mungbean-Rice Cropping System under Subtropical Climatic Conditions

Md. Khairul Alam[1], Md. Monirul Islam[2], Nazmus Salahin[1], and Mirza Hasanuzzaman[3]

[1]Soil Science Division, Bangladesh Agricultural Research Institute, Gazipur 1701, Bangladesh

[2]Tuber Crops Research Centre, Bangladesh Agricultural Research Institute, Gazipur 1701, Bangladesh

[3]Department of Agronomy, Faculty of Agriculture, Sher-e-Bangla Agricultural University, Dhaka 1207, Bangladesh

ABSTRACT

This study was conducted to know cropping cycles required to improve OM status in soil and to investigate the effects of medium-term tillage practices on soil properties and crop yields in Grey Terrace soil of Bangladesh under wheat-mungbean-T. aman cropping system. Four different tillage practices, namely, zero tillage (ZT), minimum tillage (MT), conventional tillage (CT), and deep tillage (DT), were studied in a randomized complete block (RCB) design with four replications. Tillage practices showed positive effects on soil properties and crop yields. After four cropping cycles, the highest OM accumulation, the maximum root mass density (0–15 cm soil depth), and the improved physical and chemical properties were recorded in the conservational tillage practices. Bulk and particle densities were decreased due to tillage practices, having the highest reduction of these properties and the highest increase of porosity and field capacity in zero tillage. The highest total N, P, K, and S in their available forms were recorded in zero tillage. All tillage practices showed similar yield after four years of cropping cycles. Therefore, we conclude that zero tillage with 20% residue retention was found to be suitable for soil health and achieving optimum yield under the cropping system in Grey Terrace soil (Aeric Albaquept).

INTRODUCTION

Holistic management of arable soil is the key to dealing with the most complex, dynamic, and interrelated soil properties, thereby maintaining sustainable agricultural production systems, the lone foundation of human civilization. Any management practice imposed on soil for altering the heterogenous body may result in generous or harmful outcomes [1, 2]. Unsuitable management practices cause degradation in soil health (depletion of organic matter and other nutrients) as well as decline in crop productivity [3]. Reducing disturbance of soil by reduced tillage influences several physically [4], chemically [5], and biologically [6, 7] interconnected properties of the natural body.

Soil tillage is among the important factors affecting soil properties and crop yield. Among the crop production factors, tillage contributes up to 20% [8] and affects the sustainable use of soil resources through its influence on soil properties [9]. The judicious use of tillage practices overcomes edaphic constraints, whereas inopportune tillage may cause a variety of undesirable outcomes, for example, soil structure destruction, accelerated erosion, loss of organic matter and fertility, and disruption in cycles of water, organic carbon, and plant nutrient [10]. Reducing tillage positively influences several aspects of the soil whereas excessive and unnecessary tillage operations give rise to opposite phenomena that are harmful to soil. Therefore, currently there is a significant interest and emphasis on the shift from extreme tillage to conservation and no-tillage methods for the purpose of controlling erosion process [11]. Conventional tillage practices cause change in soil structure by modifying soil bulk density and soil moisture content. In addition, repeated disturbance by conventional tillage gives birth to a finer and loose-setting soil structure while conservation and no-tillage methods leave the soil intact [12]. This difference results in a change of characteristics of the pores network. The number, size, and distribution of pores again control the ability of soil to store and diffuse air, water, and agricultural chemicals and, thus, in turn, regulate erosion, runoff, and crop performance [13]. Losses of soil organic C (SOC) and deterioration in other properties exaggerated where conventional tillage was employed [14]. With time, conservation tillage, on the other hand, improves soil quality indicators [15] including SOC storage [16].

During the first 4 years of tillage, Rhoton [17] determined a 10% loss of initial soil organic matter content with plough tillage. Mann [18] also estimated the soil organic matter depletion between 16 and 77% caused by the tillage. In most instances, increased levels of tillage or increased tillage periods resulted in reductions of soil carbon. When conventional tillage is converted to conservation tillage, both CO_2 emissions from soil and N uptake by the crop are reduced. Al-Kaisi [19] reported that reducing tillage significantly decreases SOC loss from soils with high organic matter content. Continuous cultivation for cereal cropping in the major cereal growing areas of Bangladesh leads to lowering the nutritional status of soil in most of the areas. Hence, the depletion of SOC and N concerned has taken place, a problem which needs to be managed through N fixation by the plant. In this situation, leguminous crop such as mungbean can fix N in the range of 30–40 kg N ha^{-1} [20].

Cropping system has an immense effect on physical and chemical soil properties and thereby on crop productivity [21]. Soil fertility often changes in response to land use and cropping systems and land management practices [22]. Intensive cropping promotes high levels of nutrient extraction from soils without natural replenishment. Limited practices of legume, green manure, and jute based cropping patterns have led to depletion of soil organic matter content in soils of Bangladesh [23]. Use of green manure especially legumes in a cropping pattern could help restore crop productivity. The major cereal cropping system of South Asia is rice and wheat grown on the same field but in different seasons during one year. Currently, about 12 million hectares of land in Pakistan, Nepal, India, and Bangladesh use this cropping pattern, accounting for nearly one-fourth of the region's cereal production. After rice, wheat has become an important component of cropping pattern in Bangladesh which is cultivated mostly after aman rice (lowland rice grown in the wet season from June to November in Bangladesh and east India). Crop production could be increased by adopting appropriate tillage operation and selecting suitable crops in the cropping pattern including leguminous crops, which demands intensive field research [23, 24]. Whether conservation tillage practice performs better than the long-practiced traditional tillage practices in terms of improvement of edaphic and yield influencing characters of the specific and unearth soil-water-plant ecosystem of the region is still unknown. As the conservation tillage practices have been reported to manipulate soil positively, they could also be a solution of poorly managed soil condition in the region of rice-wheat cropping system. Effect of medium-term tillage practices on soil properties in Grey Terrace soil under wheat-mungbean-T. aman (the tall traditional rice, some of which is deep water rice) cropping system has not been reported. The present study, therefore, has been initiated with the following objectives. The specific objective of the study was to observe how many cropping cycles would be required to build up organic matter (OM) in soil and the general objectives were to evaluate the effects of tillage practices on soil hydrophysical properties, to study the effect of tillage practices on the yield performances of wheat-mungbean-T. aman cropping system, and to study the medium-term effect of tillage practices on organic matter status of soil.

MATERIALS AND METHODS

Study Area

The field experiment was conducted at the Bangladesh Agricultural Research Institute (BARI), Gazipur, Bangladesh, for the four consecutive years from 2008 to 2012. The physical characteristics and chemical status of the initial soil are shown in Tables 2 and 3, respectively. The experimental site is located at the centre of the agroecological zone of Madhupur tract (AEZ-28) at about 24° 23′ north latitude and 90° 08′ east longitude having a mean elevation of 8.4 m above mean sea level. The soil belongs to the Chhiata series of the Grey Terrace soils (Aeric Albaquept) under the order Inceptisols in the USDA Soil Taxonomy [24, 25]. The morphological and taxonomical characteristics of the experimental site are shown in Table 1. The textural class was clay loam having soil pH 5.7 and the land type is medium high. Geographical position of Gazipur district is presented in Figure 1.

Table 1: Morphological and taxonomical characteristics of the experimental site

Morphological characteristics	
Locality	BARI, Gazipur, Bangladesh
Geographic position	24°-0N latitude, 90°-25E longitude, 8.40 m height above the sea level
AEZ	Madhupur tract (AEZ 28)
General soil type	Near neutral soil pH, Grey Terrace soils (Aeric Albaquept)
Taxonomic soil classification	
Order	Inceptisol
Suborder	Aquept
Subgroup	Aeric Albaquept
Soil series	Chhiata
Physiographic unit	Madhupur tract
Drainage	Moderate

Flood level	Above flood level
Vegetation	Clean cultivation and maintaining cropping pattern
Topography	Medium high land, 8.40 m height above the sea level

Table 2: Physical characteristics of the initial soil of the experimental plot

Particle size distribution	Value
Sand (%)	35.30
Silt (%)	37.29
Clay (%)	27.41
Textural class	Clay loam
Bulk density (g cm−3)	1.60
Particle density (g cm−3)	2.58
Total porosity (%)	37.98
Moisture content at field capacity (%)	24.00

Table 3: Chemical status of the initial soil of the experimental plot

Depth	pH	OM	Total N	P	S	B	Cu	Fe	Mn	Zn	K	Ca	Mg
(cm)	—	(%)	(%)	(mg kg−1)									
0–25	5.7	1.30	0.085	13	12	0.15	7.34	590	17.63	2.12	70	1202	240
Critical level			—	14	14	0.20	1.0	10.0	5.00	2.00	78	400	96

Figure 1: Geographical position of ⊙ Gazipur district (←).

The climate of the experimental area was subtropical, wet, and humid. Heavy rainfall occurs in the monsoon and is scarce in other times. The climatic data of the study area for the period from 2008 to 2012 indicates that the mean annual rainfall is above 1600 mm of which 72.2% is received during the main growing season (Kharif: one of the three seasons in Bengali crop calendar starting from mid-March and stretching to mid-October), that is, from the middle of March 2009 to the middle of October 2009. July and August alone contributed more than 50% to the annual rainfall (Figure 2). From late October to mid-March, the minimum and maximum temperatures were in the lowest range whereas from mid-March onward up to mid-October temperature was in the maximum range. However, the highest maximum temperature was recorded in May (Figure 3(a)).

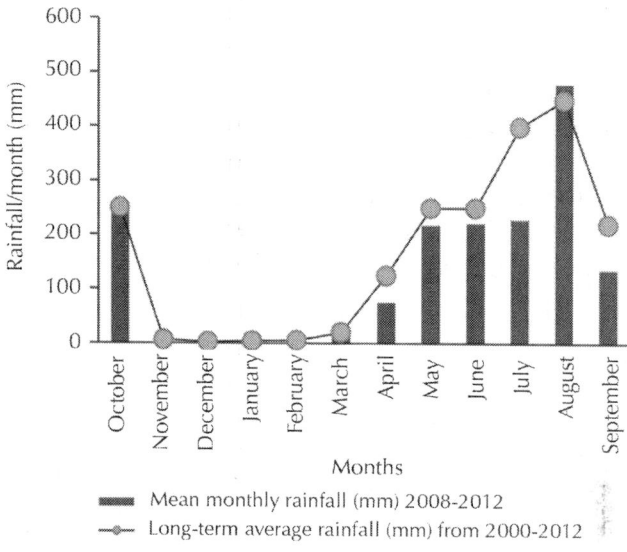

Figure 2: Rainfall (mm) distribution of the experimental site.

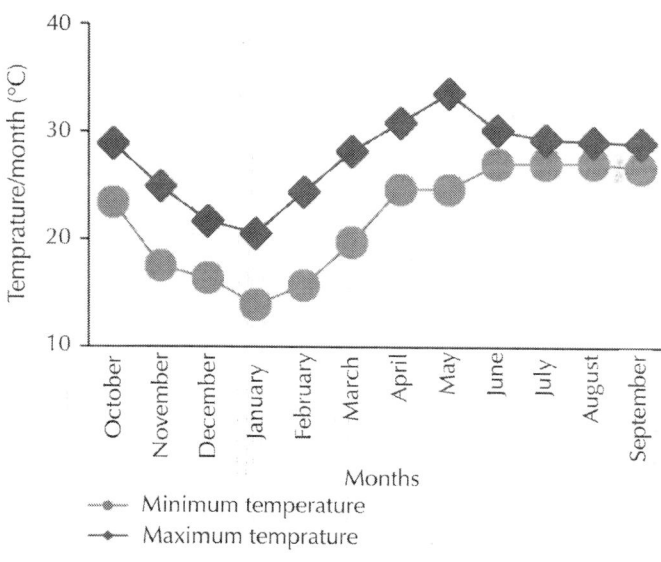

(a)

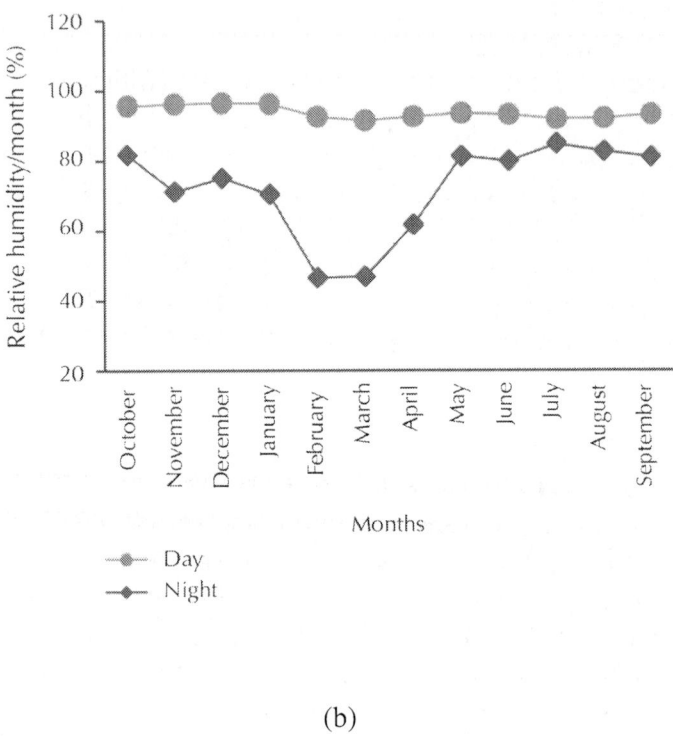

(b)

Figure 3: Temperature (a) and relative humidity (b) of the experimental site.

The periods from October to May are virtually dry. The relative humidity (%) varied between day and night of which at day time relative humidity (%) was about 90 (%) and at night it fluctuated to a wide range from 43 to 85% in February and March, respectively (Figure 3(b)).

Cropping Season

There are three major cropping seasons in Bangladesh, namely, Rabi, Kharif-I, and Kharif-II. Rabi season stretches from the middle of October to the middle of March, Kharif-I season stretches from the middle of March to the end of June, and Kharif-II season stretches from early July to the middle of October. In this experiment, wheat was grown in Rabi season, whereas mungbean and T. aman were in the Kharif-I and Kharif-II, respectively.

The Test Crop

The first crop of the cropping system was wheat (Triticum aestivum L.) cv. Sourav which was collected from the Wheat Research Centre (WRC) of BARI, Gazipur. It is a semidwarf, early maturing variety having large white grain and is suitable for cultivation in both irrigated and rain-fed conditions. The seeds of mungbean (Vigna radiata L. Wilczek) cv. BARI Mung 5 were collected from the Pulse Research Centre of BARI, Gazipur, while seeds of T. aman rice (Oryza sativa L.) cv. BRRI dhan39 were collected from the Bangladesh Rice Research Institute (BRRI), Gazipur, Bangladesh.

Experimental Design

The experiment was laid out in a randomized complete block design with four replications. The experimental design was performed as follows: zero tillage (ZT: a single slot is opened for seed sowing or transplanting), minimum tillage (MT: ploughed by power tiller maintaining depth by depth control lever up to 6–8 cm), conventional tillage (CT: similar to MT up to 14–16 cm depth), and deep tillage (DT: tillage by chisel plough up to 24–26 cm depth). The unit plot size was 5 m × 4 m.

Fertilizer Application and Other Intercultural Operations

The fertilizer doses for wheat (Sourav), mungbean, and T. aman rice were $N_{120} P_{35} K_{75} S_{20} Zn_2$, $N_{20} P_{10} K_{13} S_5$, and $N_{90} P_{18} K_{48} S_{7.5}$ kg ha^{-1} along with cow dung (CD) 5 t ha^{-1}, respectively, based on higher yield goal [25]. The fertilizer requirements were calculated on soil test basis. In the case of first crop (wheat) one third urea, whole amount of triple superphosphate (TSP) and cow dung were applied during final land preparation. The rest of the urea, MoP, gypsum, and $ZnSO_4$ were applied in two equal splits at 3rd and 5th weeks after seed sowing. For second crop (mungbean), whole amount of fertilizers was applied during final land preparation. For the third crop (T. aman rice), one third of urea and whole TSP were applied during final land preparation and the rest of the urea, MoP, gypsum, and $ZnSO_4$ were applied in two equal splits at 3rd and 5th weeks after seedling transplantation. Irrigation and other intercultural operations were done as and when necessary. The soil moisture was monitored intensively with tensiometer and sampling of soil with gravimetric method [26].

Seed Sowing/Transplanting

Wheat (cv. shatabdi) seeds were sown on the last week of November for all the years of experimentation while the first subsequent crop, mungbean (cv. BARI mung 5), was broadcasted by hands on the second week of April and the second subsequent crop, T. aman (cv. BRRI dhan 39), was transplanted on the first week of July. After picking pods twice, the total biomass of mungbean was incorporated into soil. The spacing maintained for BRRIdhan 39 and wheat was 25 × 15 cm and 15 × 5 cm, respectively. The experimental plots were kept fixed during the entire growth periods.

Sampling Procedures

In all the cropping years, the wheat was harvested in the first week of April whereas the mungbean harvesting was started in the first week of June and continued up to the third week of June. Likewise, T. aman rice was harvested in the first week of November at full maturity. Data of wheat, mungbean, and T. aman were recorded from one-square-meter area from each plot and then converted into yield per hectare. All the crops were cut at the ground level. Threshing, cleaning, and drying of grain were done separately plotwise. The weights of grain and straw were recorded plotwise. About twenty percent (20%) residue was retained in experimental field in case of wheat and rice crops. Soil samples were collected at 0–25 cm depth from each plot before sowing/planting and at the end of each cropping cycle in every year.

Soil Analyses

Soil samples were then analyzed for pH, OM, N, P, K, and Zn following standard procedures [5]. Soil pH was measured using a glass electrode pH meter (WTW pH 522) at a soil-water ratio of 1 : 2.5 as described by Ghosh [27], soil organic C was measured by Walkley and Black's wet oxidation method as described by Jackson et al. [28], and total N was measured by micro-Kjeldahl method [5]; available P was determined following the Olsen method [28], exchangeable K was determined using NH_4OAC extraction method [26], S was determined by turbidimetric method with the help of a spectrophotometer using a wave length of 420 nm [5], Ca was determined by complexometric method of titration using Na_2-TA as a complexing agent [5], Mg was determined by using NH_4OAC extraction method [26], and available Zn, Cu, Fe, and Mn were determined by using diethylenetriamine pentaacetic acid (DTPA) extraction method [29]. Particle size distribution was done by hydrometer method [26] and the textural class was determined using the USDA textural triangle. Bulk density and particle density of the soil samples were determined by core sampler method and Pycnometer method, respectively [30]. The soil porosity was computed from the relationship between bulk density and particle density using (1). Soil field capacity and permanent wilting point were measured using pressure plate apparatus, while available water content was calculated using (2) [26]. Consider

$$Porosity\,(\%)=(1-\frac{BD}{PD})\times 100, \tag{1}$$

where BD is bulk density ($g\,cm^{-3}$), PD is particle density ($g\,cm^{-3}$), and

$$d = \frac{FC-PWP}{100}\times BD \times Soil\,depth, \tag{2}$$

where d is available water content (cm) at 60 cm depth, FC is field capacity (%), and PWP is permanent wilting point (%).

The double ring infiltrometer method was used to determine the water infiltration and was computed as cumulative infiltration and rate of infiltration in $mm\,h^{-1}$.

Roots Analyses

The root mass density was measured at maximum vegetative stage in three different soil depths (0–15, 15–30, and 30–45 cm) with auger-like root sampler 15 cm (6 inch) in diameter and 22.5 cm (9 inch) in length using (3) [31]. Consider

$$\text{Root mass density} = \frac{\text{Mass of root}}{\text{Total volume of soil}} \text{mg cm}^{-3}. \qquad (3)$$

Statistical Analysis

The analysis of variance for various crop yields and soil physical and chemical properties was performed following ANOVA technique and the mean values were adjudged by Duncan's multiple range test (DMRT) method [32]. Computation and preparation of graphs were done using Microsoft Excel 2003 Program.

RESULTS

Changes of Soil Physical Properties

Bulk Density, Particle Density, Porosity, Field Capacity, and Permanent Wilting Point

Bulk density (Bd), particle density (Pd), porosity, field capacity, and permanent wilting point were influenced by the different tillage practices. Soil bulk density varied considerably among tillage practices. After four years, bulk density was decreased due to tillage practices. The highest Bd reduction (6.41%) was found in ZT followed by MT (3.95%), while DT showed the lowest reduction (Figure 4(a)). The soil particle density was decreased after four years of study. The highest decrease was noted in ZT and the minimum was in DT (Figure 4(b)). After four years of cropping cycles, porosity was increased from the initial value (6.2, 2.9, and 0.69% increase in ZT, MT, and CT, resp.) (Figure 5(a)). The field capacity (FC) was also increased due to different tillage practices. The highest FC increase (14.65%) was found in ZT followed by MT (8.52%). CT showed the lowest increase of field capacity from the first year value (Figure 5(b)). Permanent wilting point (PWP) was also influenced by the different tillage practices. After four years, the permanent wilting point was decreased due to tillage practices (Figure 6(c)). The highest reduction (11.91%) was found in ZT followed by CT (8.32%) and the lowest reduction (1.13%) in DT.

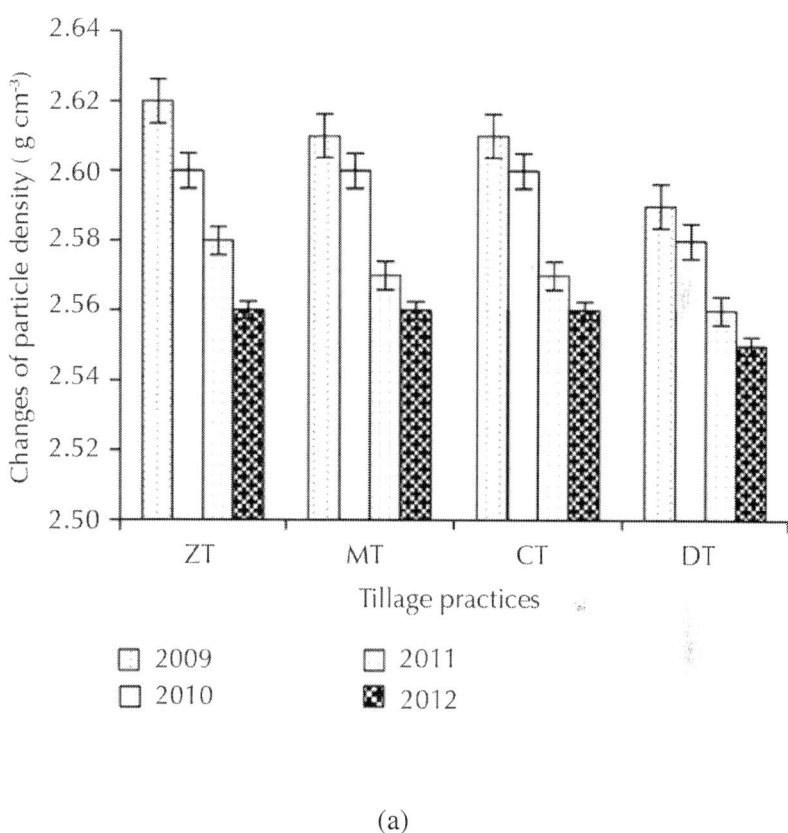

(a)

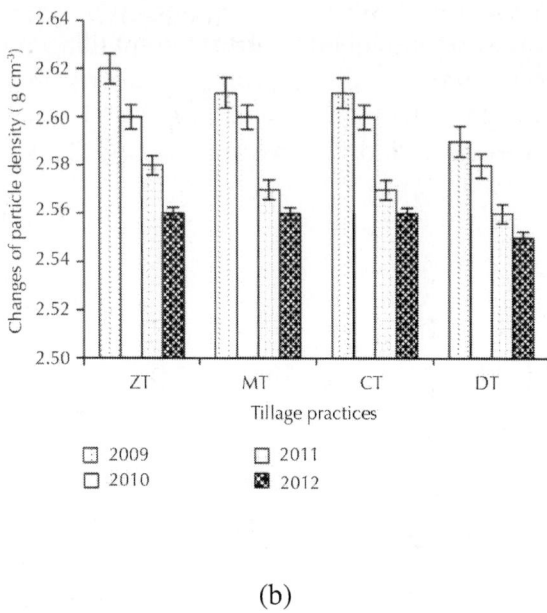

(b)

Figure 4: Change in bulk density (a) and particle density (b) as influenced by different tillage practices (most recent year first). Notes: ZT: zero tillage, MT: minimum tillage, CT: conventional tillage, and DT: deep tillage.

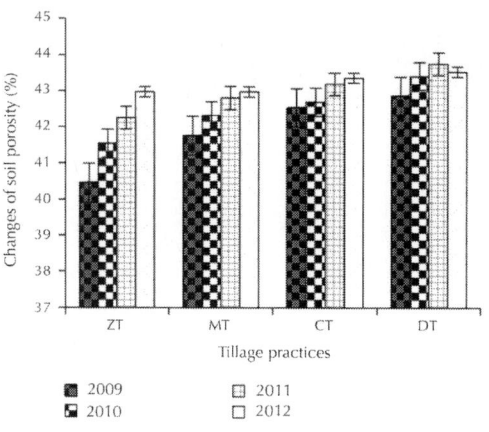

(a)

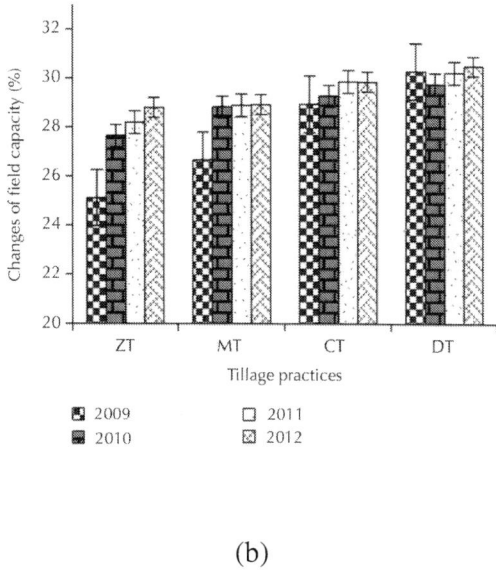

(b)

Figure 5: Change in soil porosity (a) and field capacity (b) as influenced by different tillage practices (year most recent first). Notes: ZT: zero tillage, MT: minimum tillage, CT: conventional tillage, and DT: deep tillage. Means ± SE are shown in error bar ($P \leq 0.05$).

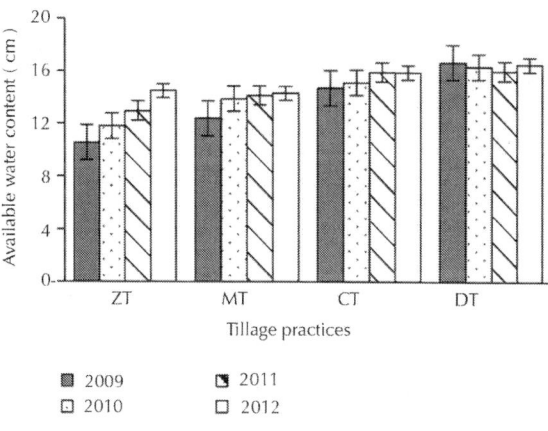

(a)

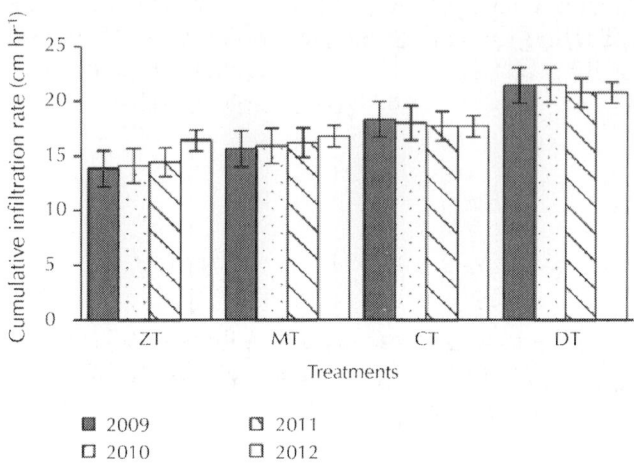

(b)

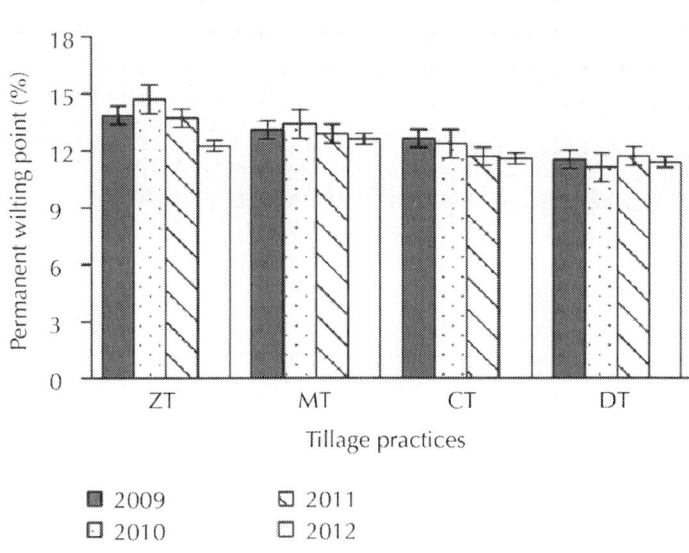

(c)

Figure 6: Effect of tillage practice on available water content of soils (a), cumulative infiltration (b), and permanent wilting point (c) (most recent year first). Notes: ZT: zero tillage, MT: minimum tillage, CT: conventional tillage, and DT: deep tillage. Means ± SE are shown in error bar ($P \leq 0.05$).

Soil Water Content

After four years of experimentation, the result showed no significant variation in available water content (AWC) due to different tillage treatments whereas AWCs were significant after completion of the first and second cropping cycles. In the end of the study, maximum available water content (AWC) was found in the deep tillage (16.50 cm) and the minimum AWC (14.30 cm) in ZT (Figure 6(a)).

Infiltration

Infiltration of water into soil was influenced by different tillage practices. The infiltration rate was found to be increased after every cropping cycle. After four years, the highest increase (18.44%) was found in ZT followed by MT (7.35%) whereas CT and DT showed decreasing trend after two years (Figure 6(b)). The maximum reduction (3.31%) was observed in DT and the minimum was in CT. The highest intercept was found in DT ($K = 5.203$) followed by CT ($K = 3.92$) which explains that deep tillage has higher initial infiltration (Figure7).

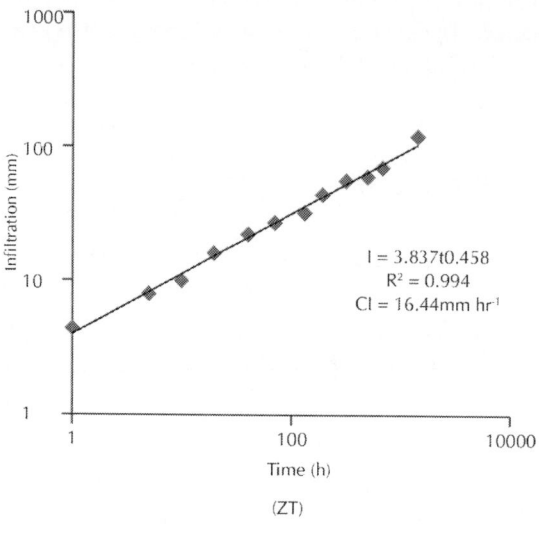

(a)

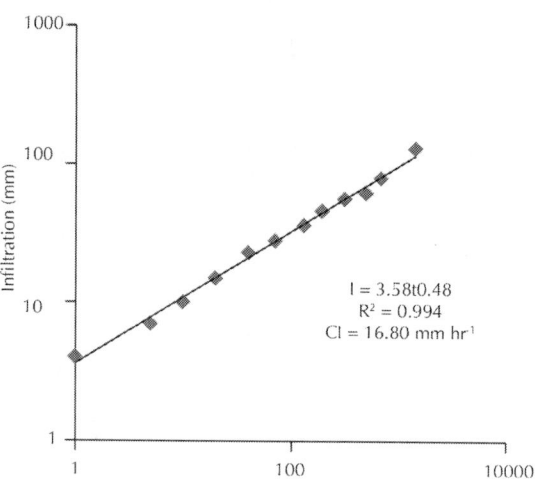

(b)

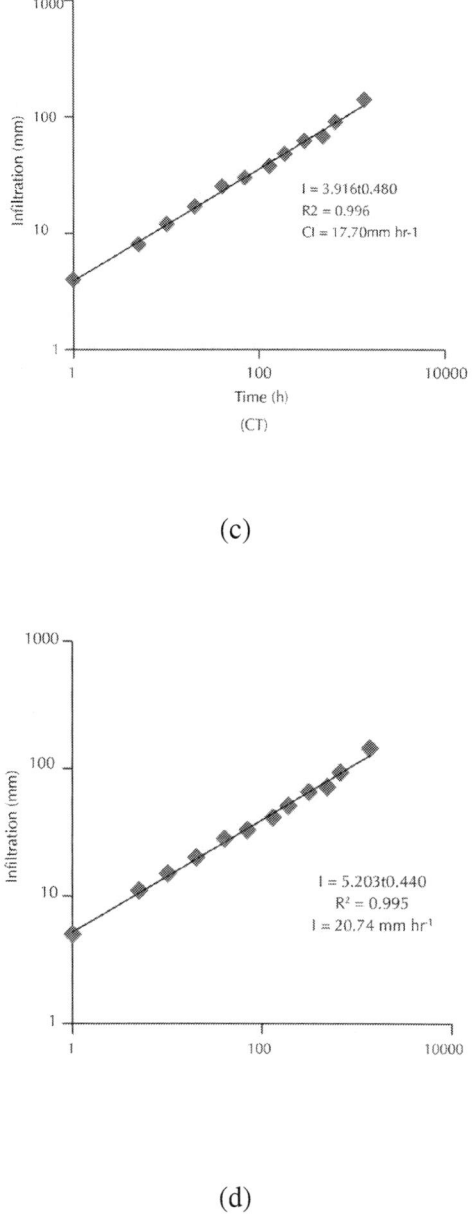

(c)

(d)

Figure 7: Cumulative infiltration (mm) over cumulative time (hour). Notes: ZT: zero tillage, MT: minimum tillage, CT: conventional tillage, and DT: deep tillage.

Organic Matter Status of Postharvest Soil

The organic matter content in the initial soil was 1.3% but changed due to different tillage practices after wheat-mungbean-T. aman cropping cycles. Organic matter ranged from 1.3 to 1.5% in 2009 and from 1.2 to 1.7% in 2010 (Figure 8(a)) of which the highest OM content of the range (1.7%) was found in ZT and the lowest (1.2%) in DT in both years. In 2011 and 2012, the maximum organic matter content (1.9 and 2.0% in 2011 and 2012, resp.) was recorded in ZT, which was followed by MT (1.8% in 2011 and 2012). DT showed the minimum organic matter (1.1%) (Figure 8(a)). In 2012, the SOM content in ZT was 34.48%, 31.03%, and 25.86% higher than the SOM in 2009, 2010, and 2011, respectively. After four years of experimentation, the SOM content in ZT was 54.76%, 32.00%, and 13.79% greater than the DT, CT, and MT, respectively (Figure 8(a)). It was found that SOM content gradually increased in ZT with increasing time but the reverse is true in the case of DT. After four years, SOM increased by 50% in ZT compared to initial status whereas MT and CT showed comparatively less increment (Figure 8(a)).

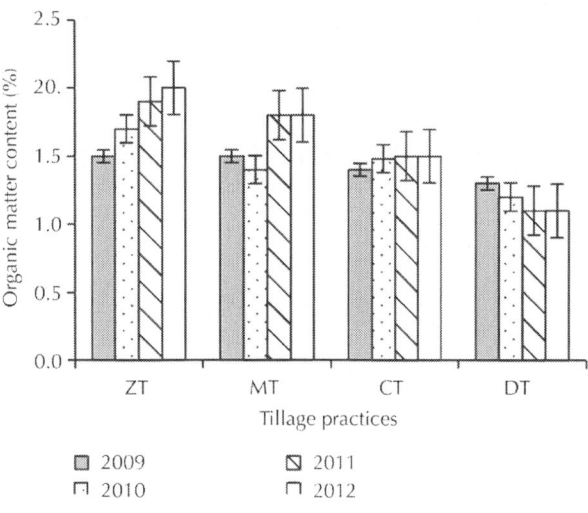

(a)

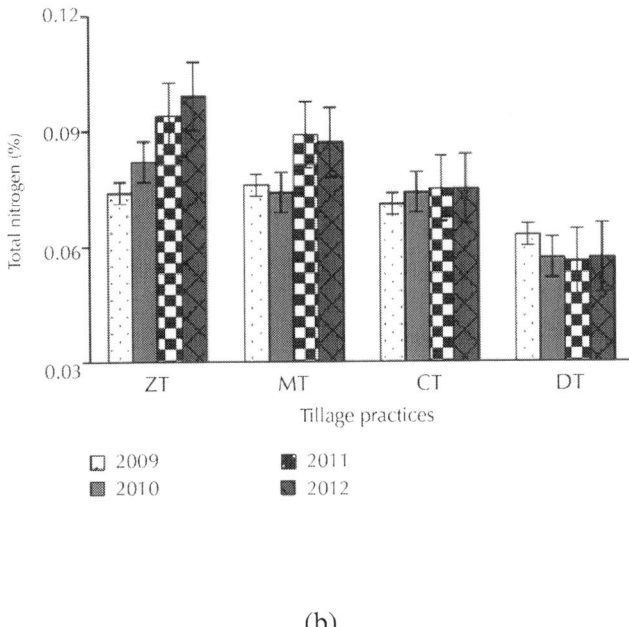

(b)

Figure 8: Change in soil organic matter (a) and total nitrogen (%) (b) due to different tillage practices (most recent year first). Notes: ZT: zero tillage, MT: minimum tillage, CT: conventional tillage, and DT: deep tillage. Means ± SE are shown in error bar ($P \leq 0.05$).

The Nutrient Status in Postharvest Soil after Every Cropping Cycle

The nutrient concentrations were significantly variable ($P \leq 0.05$) among different tillage practices (Table 8and Figure 8). The total N (%) content ranged from 0.063 to 0.076% in 2009 and from 0.057 to 0.082% in 2010. In 2010, the maximum total N content (0.082%) was found in ZT while MT showed the highest total N (0.076%) in 2009. The minimum total N content (0.063 and 0.057% for 2009 and 2010, resp.) was noted in DT (Figure 8(b)). In 2011 and 2012, ZT showed the highest total N (%) content (0.094 and 0.099% for 2011 and 2012, resp.) followed by MT and the lowest (0.056 and 0.057% for 2011 and 2012, resp.) was in DT. After four years, the total N content was 73.68,

32.0, and 13.79% higher in ZT than the DT, CT, and MT, respectively (Figure 8(b)). It was observed that the total N (%) content gradually increased in ZT and MT with progressing time (Figure 8(b)).

Phosphorus content was also significantly influenced ($P \leq 0.05$) by the different tillage practices (Table 8). In 2011 and 2012, the highest phosphorus content (18.54 and 20.32 mg kg^{-1} for 2011 and 2012, resp.) was found in ZT which was significantly higher than the other tillage practices. The lowest phosphorus content (13.76 and 14.32 mg kg^{-1}) was recorded in DT. The P content was not significantly varied ($P \leq 0.05$) among different tillage practices in 2009 and 2010. However, it ranged from 12.65 to 13.99 mg kg^{-1} and from 13.21 to 14.96 mg kg^{-1} in 2009 and 2010, respectively. The maximum P content (13.99 and 14.86 ppm for 2009 and 2010, resp.) was detected in ZT and the minimum (12.05 and 13.21 ppm for 2009 and 2010, resp.) was in DT. After four years, the available P was 41.90, 36.74, and 9.66% higher in ZT than the DT, CT, and MT, respectively (Table 8).

Sulphur content was significantly varied ($P \leq 0.05$) among different tillage practices in all the years. In 2009 and 2010, the highest sulphur (14.00 and 16.12 for 2009 and 2010, resp.) content was found in ZT followed by MT. The lowest S content (12.52 and 13.52 ppm for 2009 and 2010, resp.) was noted in DT (Table 8). In 2011 and 2012, ZT also showed the maximum S content (17.23 and 18.89 ppm for 2011 and 2012, resp.) which was significantly higher than the other tillage practices followed by MT (15.21 and 15.89 ppm for 2011 and 2012, resp.). The lowest S content (14.08 and 14.05 ppm for 2011 and 2012, resp.) was also in DT (Table 8). After four years of experimentation, available S content was 34.45, 30.73, and 18.88% higher in ZT than the DT, CT, and MT, respectively. Potassium content also followed the same trend as N, P, and S. Potassium content was significantly influenced ($P \leq 0.05$) due to different tillage practices only in 2012. It ranged from 78.0 to 93.61 ppm in 2011 and from 74.1 to 105.3 ppm in 2012. ZT showed the highest concentration of K in all the years and the minimum was in DT (Table 8). After four years of cropping cycles, available K in ZT was 42.11, 35.0, and 17.39% higher than the DT, CT, and MT, respectively (Table 8).

Effect of Tillage on Root Mass Density of Wheat

The root mass density of wheat was measured at three soil depths and variations among tillage practices at different depths (Table 4) were found. The highest root mass density was found in 0–15 cm depth followed by 15–30 cm depth. The lowest root mass density was noted in 30–45 cm depth (Table 4). In surface soil, ZT showed the maximum root mass density (9.99 mg cm^{-3}) followed by MT (9.92 mg cm^{-3}). The minimum density was recorded in DT (Table 4). In 30–45 cm depth, the highest root mass density (1.54 mg cm^{-3}) was found in DT and the lowest (0.87 mg cm^{-3}) was in ZT. As root mass density was highest in the surface soil, tillage effects on the surface would be more important than the deeper layer.

Table 4: Effect of tillage practice on root mass density of wheat and rice

Treatments	Root mass density (mg cm-3)							
	Wheat				Rice			
	0–15 cm	1 5 – 30 cm	3 0 – 45 cm	Total	0–15 cm	1 5 – 30 cm	3 0 – 45 cm	Total
ZT	9.99	1.98c	0.87b	12.84	5.40a	0.98b	0.49b	6.87
MT	9.92	2.26b	0.93b	13.11	4.90b	1.26b	0.63b	6.79
CT	8.72	2.87ab	1.21b	12.80	4.75b	1.87a	0.81a	7.43
DT	7.21	2.96a	1.54a	11.71	4.64b	1.96a	0.94a	7.54
SE (±)	0.89	0.19	0.07	—	0.09	0.18	0.06	—

Figures in a column having common letter(s) do not differ significantly at 5% level of DMRT.

ZT: zero tillage, MT: minimum tillage, CT: conventional tillage, and DT: deep tillage.

Effect of Tillage on Root Mass Density of Rice

The root mass density of rice was also significantly ($P \leq 0.05$) influenced by the tillage practices (Table 4). In the surface soil, the root mass density was significantly varied among tillage treatments. The maximum root mass density of 5.40 mg cm^{-3} was recorded in zero tillage. The deep tillage showed the highest root mass density (0.94 mg cm^{-3}) in the deeper layer and the lowest (0.49 mg cm^{-3}) was in ZT. Among the depths, surface soil showed the maximum root mass density followed by subsurface and the minimum density was noted in the deeper layer. Though the DT showed the highest root mass density in deeper layer, this layer contains very small amount of roots, whereas the maximum root mass density was found in ZT at surface soil where maximum amount of roots was recorded compared to deep layer (Table 4).

Effect of Tillage on the Yield of Wheat

The wheat yield was significantly influenced ($P \leq 0.05$) by the different tillage practices from 2009 to 2010. The highest grain yield (4.50 and 4.46 t ha^{-1} for 2009 and 2010, resp.) was found in deep tillage followed by CT (4.22 and 4.00 t ha^{-1} for 2009 and 2010, resp.). The lowest grain yield (2.76 and 3.00 t ha^{-1} for 2009 and 2010, resp.) was obtained in ZT (Table 5). The deep tillage also showed the highest straw yield (6.00 and 5.92 t ha^{-1} for 2009 and 2010, resp.) followed by CT (5.50 and 5.80 t ha^{-1} for 2009 and 2010, resp.) and MT (5.10 and 4.60 t ha^{-1} for 2009 and 2010, resp.). The minimum straw was also obtained in ZT. In 2011 and 2012, the wheat grain yield was not significantly varied ($P \leq 0.05$) among tillage practices. The wheat grain yield ranged from 3.53 to 4.13 t ha^{-1} in 2011 and from 3.69 to 4.11 t ha^{-1} in 2012. After four years, the yield gap was very minimal (negligible) among different tillage practices, though the deep tillage showed the highest yield. In the case of straw yields, a similar trend was found.

Table 5: The yield of wheat as influenced by different tillage practices

Treatment	Grain yield (t ha−1)				Straw yield (t ha−1)			
	2009	2010	2011	2012	2009	2010	2011	2012
ZT	2.76b	3.00b	3.53	3.69	4.45c	3.99	4.22	4.38
MT	3.89a	3.88a	3.71	3.78	5.10bc	4.60	4.8	4.60
CT	4.22a	4.00a	3.88	3.95	5.50ab	5.80	4.91	5.00
DT	4.50a	4.46a	4.13	4.11	6.00a	5.92	5.31	5.34
SE (±)	0.19	0.20	0.39	0.25	0.25	0.60	0.41	0.51

Figures in a column having common letter(s) do not differ significantly at 5% level of DMRT.

ZT: zero tillage, MT: minimum tillage, CT: conventional tillage, and DT: deep tillage.

Mungbean Yield

Among the four years, mungbean yield was not significantly influenced ($P \leq 0.05$) by the different tillage practices except for the yield in 2010 (Table 6). After four years (in 2012), the mungbean grain yield ranged from 792 to 820 kg ha^{-1}. The highest yield (820 kg ha^{-1}) was found in DT followed by ZT (812 kg ha^{-1}). The lowest yield (792 kg ha^{-1}) was noted in MT (Table 6). It was found that the yield difference was negligible among the tillage practices after four-year cropping cycles.

Table 6: Effect of tillage practice on the yield of mungbean

Treatment	Grain yield (kg ha−1) 2009	Biomass yield (t ha−1) 2009	Grain yield (kg ha−1) 2010	Biomass yield (t ha−1) 2010	Grain yield (kg ha−1) 2011	Biomass yield (t ha−1) 2011	Grain yield (kg ha−1) 2012	Biomass yield (t ha−1) 2012
ZT	632	6.26	644c	6.39b	780	7.26	812	7.56
MT	784	7.12	841b	6.83b	785	7.28	792	7.60
CT	837	7.82	1000a	7.68a	800	7.62	800	7.87
DT	882	8.15	1100a	8.29a	820	8.00	820	8.10
SE (±)	12.56	0.14	10.11	0.12	12.11	0.13	11.57	0.13

Figures in a column having common letter(s) do not differ significantly at 5% level of DMRT.

ZT: zero tillage, MT: minimum tillage, CT: conventional tillage, and DT: deep tillage.

Effect of Tillage Practices on the Yield of T. *aman*

In 2009 and 2010, the T. aman yields were significantly influenced ($P \leq 0.05$) by the different tillage practices. The grain yield ranged from 2.87 to $4.56 \, t \, ha^{-1}$ in 2009 and from 3.64 to $4.63 \, t \, ha^{-1}$ in 2010. The highest grain yield (4.50 and $4.63 \, t \, ha^{-1}$ for 2009 and 2010, resp.) was found in DT. The minimum grain yield (2.87 and 3.64 for 2009 and 2010, resp.) was recorded in ZT. After two years, rice yield was not significantly variable ($P \leq 0.05$) among tillage practices (Table 7).

Table 7: The yield of T. *aman* as influenced by different tillage practices

Treatment	Grain yield (t ha−1)				Straw yield (t ha−1)			
	2009	2010	2011	2012	2009	2010	2011	2012
ZT	2.87c	3.64b	4.18	4.49	3.07b	3.75	3.95	4.54
MT	3.77b	3.84ab	4.24	4.30	4.13a	3.96	3.96	4.32
CT	4.40a	4.29ab	4.29	4.37	4.68a	4.39	4.40	4.40
DT	4.56a	4.63a	4.43	4.51	4.84a	4.69	4.60	4.60
SE (±)	0.13	0.28	0.17	0.12	0.27	0.31	0.52	0.11

Figures in a column having common letter(s) do not differ significantly at 5% level of DMRT.

ZT: zero tillage, MT: minimum tillage, CT: conventional tillage, and DT: deep tillage.

Table 8: Effect of tillage practice on available P, available K, and available S after wheat-mungbean-T. *aman* cropping sequence

Treatment	P (ppm)				K (ppm)				S (ppm)			
	2009	2010	2011	2012	2009	2010	2011	2012	2009	2010	2011	2012
ZT	13.99a	14.96a	18.54a	20.32a	89.7	89.7	93.6	105.3a	14.00a	16.12a	17.23a	18.89a
MT	13.97a	14.13a	15.43b	18.53b	89.7	85.8	81.9	89.7b	13.56ab	15.27ab	15.21b	15.89b
CT	13.21a	13.92a	14.90b	14.86c	85.8	78.0	81.9	78.0c	13.32ab	14.23bc	14.43b	14.45b
DT	12.65a	13.21a	13.76b	14.32c	85.8	78.0	78.0	74.1c	12.52b	13.52c	14.08b	14.05b
SE (±)	0.95	1.09	0.61	0.27	8.08	5.69	5.39	2.06	0.36	0.43	0.44	0.65

Figures in a column having common letter(s) do not differ significantly at 5% level of DMRT.

ZT: zero tillage, MT: minimum tillage, CT: conventional tillage, and DT: deep tillage.

DISCUSSION

The experiment was conducted during four years indicating that zero and minimum tillage practices had significant influence on soil physical and chemical properties and yields of crops in wheat-mungbean-T. amancropping system, compared to conventional and deep tillage practices.

The bulk density (Bd) was decreased by 9.59, 9.59, 10.34, and 11.11% in DT, CT, MT, and ZT, respectively, compared to initial value. Different tillage practices showed more or less similar influence on bulk density. After the completion of four years, it was found that there was no significant ($P \leq 0.05$) difference among the different tillage practices. This might be due to the deposition of OM in ZT practice. Soil bulk density is the significant indicator of change of soil physical health and water retention capacity under different tillage depths [33]. A similar result was reported by Sarwar et al. [34]. In New South Wales (NSW), Australia, the soil Bd was reduced by 6.7% in no tillage (at 50 cm depth) compared to conventional tillage after 14 years [35]. He et al. [35] reported that the mean bulk density (in 0–30 cm soil layer depth) under NT and CT treatments was 1.40 and 1.41 Mg m^{-3}, respectively, and the difference was negligible in the long terms which is in agreement with the findings of our study. In Chinese Loess Plateau, crop stubble retention under no tillage and controlled traffic has been reported to increase soil organic matter and biotic activity, thereby reducing bulk density in the surface soil layer [35, 36]. Soil organic C has a direct impact on the bulk density or inversely on the porosity of soil, since the particle density of organic matter is considerably lower than that of mineral soil and soil organic matter is often associated with increased aggregation and permanent pore development as a result of soil biological activity [37]. The changes in soil bulk density in 0–0.30 m soil layer are consistent with the porosity results. After 8 years of different management, the mean soil bulk density in 2007 was 0.8–1.5% lower in NT than the CT at Daxing and Changping. The reduced bulk density in NT could be attributed to higher organic matter content [38] and better aggregation [39].

Soil particle density (Pd) was varied insignificantly ($P \leq 0.05$) among tillage practices after four years of experimentation. For an average between 0 and 25 cm depth, particle density was found to be decreasing in ZT with time ($2.56\,g\,cm^{-3}$) as compared to deep tillage ($2.55\,g\,cm^{-3}$) where particle density was stuck at a fairly constant level. The decrease of Pd in ZT ($P \leq 0.05$) might be due to accumulation of organic matter (OM) with time. A similar result was also observed by Rühlmann et al. [39] where Pd decreased in top soil and it was related to variation in SOC.

The effects of tillage practices on porosity were smaller but consistently positive over years. After 4 years, porosity was increased in soil from the initial year due to tillage practices. The increase of soil porosity in ZT might be due to the addition of OM and crop residues which was caused by zero and minimum disturbance of soil. Similar results were also reported by He et al. [35]. Many studies have indicated that tillage systems significantly influenced the soil pore size distribution [40]. However, our results were in agreement with the findings of He et al. [35] who reported that total porosity in the 0–15 cm layer was similar under different treatments and also with that of the work of Zhang and Song et al. [40] where no significant difference in porosity was found at the surface layer. Zhu et al. [41] also reported that no tillage with mulch was found to increase 5.5% of total porosity in 0–30 cm soil layer compared to traditional tillage after 4 years.

Soil moisture retentive characteristics (SWR) were varied by tillage practices. Soil water retention (SWR) at field capacity (−33 kPa) was initially higher in the deep tillage ($P \leq 0.05$) but it was gradually increasing in the soil treated with zero tillage with the advancement of experiment. Plant-available water content gave almost similar result ($P \leq 0.05$) to soil water retention at FC. Higher difference was also observed in 0–25 cm depth after completion of the first cropping cycle, where AWC was 36.68, 28.18, and 14.78% lower in ZT than the DT, CT, and MT, respectively. After four cropping cycles, the results reflected insignificant AWC among different tillage practices. The increasing trend of water retention in the soil under ZT practices also implied water uptake increase by the crop, resulting in a gradual improvement of crops yield in zero tillage compared to the other tillage practices in the dry season where yields almost remained constant or decreasing in some cases with time. Soils under no-tillage practices have greater water storage capacity than the tilled soils [42]. Fernández-Ugalde [43] conducted an experiment for a medium-term basis and found that the SWR at field capacity was significantly higher in NT than the CT and reported that these differences were particularly noticeable in the soil surface depth where water retention was 23% lower in CT than in NT. In the present study, SWR was found to be increasing in ZT practice with experiment progressing ahead even though the soil water content at field capacity (FC) during the initial year was found to be significantly higher ($P \leq 0.05$) with DT. In the long run, SWR and AWC would be found to be significantly higher in zero tillage than the other tillage practices as the experiment showed the evidence of OM build-up and other physical characteristics favourable for this. Besides, infiltration is an important soil feature controlling leaching, runoff, and crop water availability [44].

After the first cropping cycle, the variation in soil water infiltration was higher among different tillage practices than the infiltration variation four years apart which was found to be narrowing down ($P \leq 0.05$). ZT practices promote infiltration and water retention year after year. Schwen et al. [44] reported that soils under no-tillage treatment have greater infiltration rates than the tilled soils. With management for less than a few years, water infiltration in NT may be similar or lower than the CT due to initial compaction and lack of sufficient biological activity for development of stable soil structure [45]. Conservation tillage practice with judicious crop residue management improves aggregate stability [46] and leads to reduced soil detachment and improved infiltration rates [47]. Surface OM is also essential for water infiltration and conservation of nutrients [48]. Wang et al. [49] also reported that conservation tillage may delay run-off by 12–16 min in heavy rainfall and improve final infiltration rate by 60.9% in comparison with conventional mouldboard ploughing in Shanxi province.

The root mass density (RMD) of wheat and rice varied ($P \leq 0.05$) among the tillage practices and different soil depths. The total RMD of three sampling depths for both crops was found close in range (13.11–11.71 and 7.54–6.87 mg cm^{-3} for wheat and rice, resp.). In 0–15 cm depth, the roots growth was higher in ZT and MT than the CT and DT but the reverse is true in case of subsurface (15–30 cm) and deep soils (30–45 cm). Therefore, ZT plays an important role in root mass density distribution in the soil. The incorporation of biomass from mungbean favoured maximum roots growth [50, 51]. The root mass density was drastically reduced downward, which was associated with the increased soil bulk density in deeper zone. Root proliferation or extensibility was obstructed by the dense or compact layer of the soil profile [52]. Similar results were found by Parker and Lear [53] and Alam and Matin [54] in different crops.

It was observed that the OM content (%) was found to be decreased in deep tillage after each cropping cycle of wheat-mungbean-T. aman whereas organic matter was gradually deposited in the soils where no or minimum disturbance occurred throughout the four cropping cycles. A similar result was also found by Chan and Heenan et al. [55] in different tillage practices. Zero tillage along with addition of organic matter and crop residues in the cropping systems has been reported to increase soil organic matter significantly in the 0–25 cm soil layer compared to DT after 4 years. Zhu et al. [41] also observed a similar result where ZT had 4.3% SOM in the 0–30 cm soil layer compared to traditional tillage after 4 years. In addition, improvements of crop yields have been documented where conservation tillage was practiced [56, 57]. Ma and Tong [57] reported that the winter wheat yield in conservation tillage was 10–20% higher than the conventional tillage in Shandong, northern China. Mean wheat yield improvement in no tillage was estimated to be 4.3% between 2003 and 2004 in the more arid Hexi Corridor area of northwest China [58]. In central Texas, United States, after twenty years in wheat cropping system, soil organic matter and total N were increased by 28 and 33% in no tillage at 0–15 cm soil depth [59]. Conservation tillage was also showed to improve soil water content and crop yields in many environments [7, 60], whereas Hammel et al. [60] reported negative effects of no tillage on crop yields in arid areas of the United States. However, frequent and excessive tillage and residue removal in CT and deep tillage by chiseling resulted in significant loss of SOM [61]. Tillage-induced changes in soil organic N are often directly related to changes in SOC. ZT and MT showed significantly ($P \leq 0.05$) higher concentrations of available N in the surface soil. Soil available P was also significantly ($P \leq 0.05$) improved by the MT and ZT, particularly in 0–25 cm soil depth. The accumulation of P at the topsoil in ZT and MT can be explained by the limited downward movement of particle-bound P in no-till and minimum-till soils and the upward movement of nutrients from deeper layers through uptake by roots [62]. Roldan et al. [62] observed that SOM increased by up to 15% through no tillage and minimum tillage at 0–50 mm soil depth in Mexico. The significant increases of available N and P in conservation tillage practices were also consistent with the findings of other researchers [63, 64]. In a study, Reyes et al. [64]

reported that soil organic carbon (SOC) was higher in NT (2.77% in 0–15 cm depth) compared to CT (2.22% in 0–15 cm depth). Reicosky et al. [65] also reported that SOM content was increased under conservation tillage practices following the accumulation rate from 0 to 1.15 t C ha^{-1} yr^{-1} with the highest values in temperate climatic condition. Similar data were also observed by Lal et al. [66] where organic carbon accumulation rate ranged from 0.1 to 0.5 t ha^{-1} yr^{-1}. This aspect is very important due to the multiple roles played by the organic matter in the soil. It regulates biological, physical, and chemical processes that collectively determine soil health.

After four years of experimentation, it was found that there was no difference in grain yield of rice as influenced by DT and ZT. This might be due to the build-up of organic matter in the zero tillage practice which occurred with the progress of cropping cycles. In the present study, the improved soil chemical and physical properties were probably responsible for the increased crop yields in conservation tillage practices (ZT and MT) in Grey Terrace soil under wheat-mungbean-T. aman cropping system. As reported by Liao et al. [67] and Xue et al. [68], conservation tillage practices have been shown to increase crop yield considerably.

CONCLUSIONS

After four years, different tillage practices showed that they influenced soil physical and chemical properties along with the improvement of SOM status under wheat-mungbean-T. aman cropping systems. ZT with mungbean biomass and residue incorporation conserved moisture in the soil profile and improved other soil properties, reduced the bulk density, and increased OM, porosity, AWC, and RMD. After four years, the chemical properties were also improved due to ZT and MT practices. The highest total N (%), P, K, and S in their available forms were found in zero tillage. All tillage practices showed statistically similar yield after four years of cropping cycles. Therefore, zero tillage (minimum soil disturbance) with 20% residue retention was found to be suitable to improve soil conditions and to achieve optimum yield under wheat-mungbean-T.aman cropping system in the Grey Terrace soil (Aeric Albaquept).

ACKNOWLEDGMENTS

The authors are thankful to the Ministry of Agriculture, Peoples' Republic of Bangladesh, for financial support during the study period. They are grateful to BARC (Bangladesh Agricultural Research Council) for the advice. Special thanks are also due to Dr. M.D. Azizul Haque and Dr. Rowshan Ara Begum and their team for physical and chemical laboratory facilities.

REFERENCES

1. R. Derpsch, T. Friedrich, A. Kassam, and L. Hongwen, "Current status of adoption of no-till farming in the world and some of its main benefits," International Journal of Agricultural and Biological Engineering, vol. 3, no. 1, pp. 1–25, 2010.

2. F. Wolfarth, S. Schrader, E. Oldenburg, J. Weinert, and J. Brunotte, "Earthworms promote the reduction of Fusarium biomass and deoxynivalenol content in wheat straw under field conditions," Soil Biology and Biochemistry, vol. 43, no. 9, pp. 1858–1865, 2011.

3. M. E. Ramos, A. B. Robles, A. Sánchez-Navarro, and J. L. González-Rebollar, "Soil responses to different management practices in rainfed orchards in semiarid environments," Soil and Tillage Research, vol. 112, no. 1, pp. 85–91, 2011.

4. R. López-Garrido, M. Deurer, E. Madejón, J. M. Murillo, and F. Moreno, "Tillage influence on biophysical soil properties: the example of a long-term tillage experiment under Mediterranean rainfed conditions in South Spain," Soil and Tillage Research, vol. 118, pp. 52–60, 2012.

5. A. L. Page, R. H. Miller, and D. R. Kuny, Methods of Soil Analysis. Part 2, American Society of Agronomy, Soil Science Society of America, Madison, Wis, USA, 2nd edition, 1989.

6. C. J. Bronick and R. Lal, "Soil structure and management: a review," Geoderma, vol. 124, no. 1-2, pp. 3–22, 2005.

7. A. Muñoz, A. López-Piñeiro, and M. Ramírez, "Soil quality attributes of conservation management regimes in a semi-arid region of south western Spain," Soil and Tillage Research, vol. 95, no. 1-2, pp. 255–265, 2007.

8. K. Khurshid, M. Iqbal, M. S. Arif, and A. Nawaz, "Effect of tillage and mulch on soil physical properties and growth of maize," International Journal of Agriculture and Biology, vol. 8, pp. 593–596, 2006.

9. R. Lal and B. A. Stewart, Eds., Principles of Sustainable Soil Management in Agroecosystems, vol. 20, CRC Press, 2013.

10. R. Lal, "Tillage effects on soil degradation, soil resilience, soil quality, and sustainability," Soil and Tillage Research, vol. 27, no. 1–4, pp. 1–8, 1993.

11. M. Iqbal, A. U. Hassan, A. Ali, and M. Rizwanullah, "Residual effect of tillage and farm manure on some soil physical properties and growth of wheat (Triticum aestivum L.)," International Journal of Agriculture and Biology, vol. 1, pp. 54–57, 2005.

12. M. Rashidi and F. Keshavarzpour, "Effect of different tillage methods on grain yield and yield components of maize (Zea mays L.)," International Journal of Rural Development, vol. 2, pp. 274–277, 2007.

13. F. U. H. Khan, A. R. Tahir, and I. J. Yule, "Intrinsic implication of different tillage practices on soil penetration resistance and crop growth," International Journal of Agriculture and Biology, vol. 1, pp. 23–26, 2001.

14. D. S. Powlson, A. Bhogal, B. J. Chambers et al., "The potential to increase soil carbon stocks through reduced tillage or organic material additions in England and Wales: a case study," Agriculture, Ecosystems and Environment, vol. 146, no. 1, pp. 23–33, 2012.

15. C. Plaza, D. Courtier-Murias, J. M. Fernández, A. Polo, and A. J. Simpson, "Physical, chemical, and biochemical mechanisms of soil organic matter stabilization under conservation tillage systems: a central role for microbes and microbial by-products in C sequestration," Soil Biology and Biochemistry, vol. 57, pp. 124–134, 2013.

16. K. L. Sharma, J. K. Grace, R. Milakh et al., "Improvement and assessment of soil quality under long-term conservation agricultural practices in hot, arid tropical aridisol," Communications in Soil Science and Plant Analysis, vol. 44, no. 6, pp. 1033–1055, 2013.

17. F. E. Rhoton, "Influence of time on soil response to no-till practices," Soil Science Society of America Journal, vol. 64, no. 2, pp. 700–709, 2000.

18. L. K. Mann, "Changes in soil carbon storage after cultivation," Soil Science, vol. 142, pp. 279–288, 1986.

19. M. Al-Kaisi, "Impact of tillage and crop rotation systems on soil carbon sequestration," Iowa State University, Ames, Iowa, USA, 2001.

20. P. Shah, K. R. Dahal, S. K. Shah, and D. R. Dangol, "Effect of tillage, mulch, and time of nitrogen application on the yield of wheat (Triticum aestivum L.)," Agriculture Development Journal, vol. 8, pp. 9–19, 2011.

21. M. M. Rahman and S. L. Ranamukhaarachchi, "Fertility status and possible environmental consequences of Tista Floodplain soils in Bangladesh," Thammasaltn International Journal of Science and Technology, vol. 8, no. 3, pp. 11–19, 2003.

22. BARC (Bangladesh Agricultural Research Council), Agricultural Research Priority:Vision 2030 and beyond—a final Report by M. Jahiruddin (Professor, Department of Soil Science, Bangladesh Agricultural University) and M. A. Satter (Chief Scientific Officer, Land and Soil Resource Management, BARC, Bangladesh). Sub sector: Land and Soil Resource Management Bangladesh Agril. Res. Council, Farmgate, Dhaka, 2010.

23. M. K. Alam, Effect of tillage depths and cropping patterns on soil properties and crop productivity [M.S. thesis], Department of Soil Science, Bangabandhu Sheikh Mujibur Rahman Agricultural University, 2010.

24. SRDI, "Bhumi and Mrittika Sampad Bhavohar Nirdeshika, Gazipur Sadar Upazila, Gazipur Zilla," Soil Resource Development Institute. Ministry of Agriculture, Government of the Peoples Republic of Bangladesh, 2005.

25. BARC (Bangladesh Agricultural Research Council), Fertilizer Recommendation Guide-2005, vol. 45 ofSoils Pub., Bangladesh Agricultural Research Council, Farmgate, Bangladesh, 2005.

26. C. A. Black, Method of Soil Analysis Part-I and II, American Society of Agronomy, Madison, Wis, USA, 1965.

27. P. Ghosh, "Institute of agriculture, Visva-Bharati, Srinike tan-731-236. West Bengal, India," Indian Journal of Agricultural Research, vol. 32, pp. 75–80, 1983.

28. M. L. Jackson, Soil Chemical Analysis, Constable and Co. Ltd. Prentice Hall of India Pvt. Ltd, New Delhi, India, 1973.

29. W. L. Lindsay and W. A. Norvell, "Development of a DTPA test for zinc, iron, manganese, and copper," Soil Science Society of American Journal, vol. 42, pp. 421–428, 1978.

30. Z. Karim, S. M. Rahman, M. I. Ali, and A. J. M. S. Karim, Soil Bulk Density. A Manual for Determination of Soil Physical Parameters, Soils and Irrigation Division, BARC, 1988.

31. J. J. Schuurman and M. A. J. Goodewaagen, Methods for the Examination of Root Systems and Roots, Centre of Agricultural Publishing and Documentation, Wageningen, The Netherlands, 2nd edition, 1971.

32. R. C. B. Steel and J. H. Torii, Principles and Procedures of Statistics, Mc Graw Hall, New York, NY, USA, 1960.

33. H. Jin, L. Hongwen, W. Xiaoyan et al., "The adoption of annual subsoiling as conservation tillage in dryland maize and wheat cultivation in northern China," Soil and Tillage Research, vol. 94, no. 2, pp. 493–502, 2007.

34. G. Sarwar, H. Schmeisky, N. Hussain, S. Muhammad, M. Ibrahim, and E. Safdar, "Improvement of soil physical and chemical properties with compost application in rice-wheat cropping system," Pakistan Journal of Botany, vol. 40, no. 1, pp. 275–282, 2008.

35. J. He, Q. Wang, H. Li et al., "Soil physical properties and infiltration after long-term no-tillage and ploughing on the Chinese Loess Plateau," New Zealand Journal of Crop and Horticultural Science, vol. 37, no. 3, pp. 157–166, 2009.

36. A. J. Franzluebbers, J. A. Stuedemann, H. H. Schomberg, and S. R. Wilkinson, "Soil organic C and N pools under long-term pasture management in the Southern Piedmont USA," Soil Biology and Biochemistry, vol. 32, no. 4, pp. 469–478, 2000.

37. B. S. Brar, K. Singh, and G. S. Dheri, "Carbon sequestration and soil carbon pools in a rice-wheat cropping system: effect of long-term use of inorganic fertilizers and organic manure," Soil & Tillage Research, vol. 128, pp. 30–36, 2013.

38. Z. Yang, L. Zhou, Y. Lv, H. Li, D. Sun, and M. Yu, "Soil aggregates features under different tillage systems in North China plain," Advanced Science Letters, vol. 19, no. 9, pp. 2761–2766, 2013.

39. J. Rühlmann, M. Körschens, and J. Graefe, "A new approach to calculate the particle density of soils considering properties of the soil organic matter and the mineral matrix," Geoderma, vol. 130, no. 3-4, pp. 272–283, 2006.

40. S. B. Zhang and C. C. Song, "Effects of different land-use on soil physical-chemical properties in the Sanjiang Plain," Chinese Journal of Soil Science, vol. 3, pp. 25–30, 2004 (Chinese).

41. B. Y. Zhu, J. H. Huang, Y. Y. Huang, and J. Liu, "Effect of continuous tillage-free practice on the grain yield of semilate rice and soil physicochemical property," Fujian Journal of Agricultural Science, vol. 14, pp. 159–163, 1999.

42. O. Fernández-Ugalde, I. Virto, P. Bescansa, M. J. Imaz, A. Enrique, and D. L. Karlen, "No-tillage improvement of soil physical quality in calcareous, degradation-prone, semiarid soils," Soil and Tillage Research, vol. 106, no. 1, pp. 29–35, 2009.

43. A. J. Fernández-Ugalde, "Water infiltration and soil structure related to organic matter and its stratification with depth," Soil and Tillage Research, vol. 66, no. 2, pp. 197–205, 2002.

44. A. Schwen, G. Bodner, P. Scholl, G. D. Buchan, and W. Loiskandl, "Temporal dynamics of soil hydraulic properties and the water-conducting porosity under different tillage," Soil and Tillage Research, vol. 113, no. 2, pp. 89–98, 2011.

45. P. W. Unger, "Infiltration of simulated rainfall: tillage system and crop residue effects," Soil Science Society of American Journal, vol. 56, no. 1, pp. 283–289, 1992.

46. R. Sonnleitner, E. Lorbeer, and F. Schinner, "Effects of straw, vegetable oil and whey on physical and microbiological properties of a chernozem," Applied Soil Ecology, vol. 22, no. 3, pp. 195–204, 2003.

47. E. T. Ekwue, "Quantification of the effect of peat on soil detachment by rainfall," Soil and Tillage Research, vol. 23, no. 1-2, pp. 141–151, 1992.

48. A. J. Franzluebbers, "Soil organic matter stratification ratio as an indicator of soil quality," Soil and Tillage Research, vol. 66, no. 2, pp. 95–106, 2002.

49. X. Y. Wang, H. W. Gao, B. Du, and H. W. Li, "Conservation tillage effect on runoff and infiltration under simulated rainfall," Journal of Soil Water Conservation, vol. 3, pp. 23–25, 2000 (Chinese).

50. N. N. Kulikova, A. N. Suturin, A. M. Antonenko, S. M. Boiko, and L. F. Paradina, "Organomineral composts from the wastes of the pulp and paper industry and their effect on soil fertility," Eurasian Soil Science, vol. 29, no. 7, pp. 836–840, 1996.

51. U. K. Mandal, G. Singh, U. S. Victor, and K. L. Sharma, "Green manuring: its effect on soil properties and crop growth under rice—Wheat cropping system," European Journal of Agronomy, vol. 19, no. 2, pp. 225–237, 2003.

52. A. E. Hassan, Y. Kitamura, E. S. Ahmed, G. Samir, and M. Irshad, "Effect of irrigation schedules and tillage systems on rice productivity and soil physical characteristics—a case study in the north Nile Delta, Egypt," in Proceedings of the 19th International Congress on Irrigation and Drainage, Beijing, China, September 2005.

53. M. M. Parker and D. H. van Lear, "Soil heterogeneity and root distribution of mature loblolly pine stands in Piedmont soils," Soil Science Society of America Journal, vol. 60, no. 6, pp. 1920–1925, 1996.

54. S. M. K. Alam and M. A. Matin, "Impact of tillage and root growth of yield of rice in a Silt Loam soil," Journal of Biological Sciences, vol. 2, pp. 548–550, 2002.

55. K. Y. Chan and D. P. Heenan, "The effects of stubble burning and tillage on soil carbon sequestration and crop productivity in south eastern Australia," Soil Use and Management, vol. 21, pp. 427–431, 2005.

56. J. Wu, Z. L. Zhong, J. G. Zheng, and X. L. Jiang, "Influences of residue mulching treatment on soil physical and chemical properties and crop yields in SW China," Journal of Agricultural Science, vol. 2, pp. 192–195, 2006 (Chinese).

57. G. Z. Ma and H. Tong, "The study and analyzing of the affection to the winter wheat by conservation agricultural technology in the dry land farming area," Agricultural Mechanization Research, vol. 5, pp. 139–142, 2007.

58. A. L. Wright, F. Dou, and F. M. Hons, "Soil organic C and N distribution for wheat cropping systems after 20 years of conservation tillage in central Texas," Agriculture, Ecosystems and Environment, vol. 121, no. 4, pp. 376–382, 2007.

59. A. Hemmat and I. Eskandari, "Conservation tillage practices for winter wheat-fallow farming in the temperate continental climate of northwestern Iran," Field Crops Research, vol. 89, no. 1, pp. 123–133, 2004.

60. J. E. Hammel, "Long-term tillage and crop rotation effects on winter wheat production in Northern Idaho," Agronomy Journal, vol. 87, no. 1, pp. 16–22, 1995.

61. A. M. Urioste, G. G. Hevia, E. N. Hepper, L. E. Anton, A. A. Bono, and D. E. Buschiazzo, "Cultivation effects on the distribution of organic carbon, total nitrogen and phosphorus in soils of the semiarid region of Argentinian Pampas," Geoderma, vol. 136, no. 3-4, pp. 621–630, 2006.

62. A. Roldan, J. R. Salinas-García, M. M. Alguacil, E. Díaz, and F. Caravaca, "Soil enzyme activities suggest advantages of conservation tillage practices in sorghum cultivation under subtropical conditions,"Geoderma, vol. 129, no. 3-4, pp. 178–185, 2005.

63. G. A. Thomas, R. C. Dalal, and J. Standley, "No-till effects on organic matter, pH, cation exchange capacity and nutrient distribution in a Luvisol in the semi-arid subtropics," Soil and Tillage Research, vol. 94, no. 2, pp. 295–304, 2007.

64. J. I. Reyes, P. Silva, E. Martínez, and E. Acevedo, "Labranzay propiedades de un suelo aluvial de Chile Central," in Proceedings del IX Congreso Nacional de la Ciencia del Suelo, pp. 78–81, Sociedad Nacional de la Ciencia del Suelo, Talca, Chile, 2002.

65. D. C. Reicosky, W. D. Kemper, G. W. Langdale, C. L. Douglas, and P. E. Rasmussen, "Soil organic matter changes resulting from tillage and biomass production," Journal of Soil and Water Conservation, vol. 50, no. 3, pp. 253–261, 1995.

66. R. Lal, J. M. Kimble, R. F. Follett, and C. V. Cole, The Potential of US Cropland to Sequester Carbon and Mitigate the Greenhouse Effect, Ann Arbor Press, Chelsea, Mich, USA, 1998.

67. Y. C. Liao, S. M. Han, and X. X. Wen, "Soil water content and crop yield effects of mechanized conservative tillage-cultivation system for dryland winter wheat in the loess tableland," Transactions of the Chinese Society of Agricultural Engineering, vol. 18, no. 4, pp. 68–71, 2002 (Chinese).

68. S. P. Xue, Q. Yan, R. X. Zhu, H. Q. Wang, and W. S. Yao, "Experimental study on conservation tillage technology of complete corn stalk cover by mechanization and furrow sowing beside film mulching,"Transactions of Chinese Science, Agriculture and Engineering, vol. 7, pp. 81–83, 2005 (Chinese).

10

Advances in Agronomic Management of Indian Mustard (Brassica juncea (L.) Czernj. Cosson): An Overview

Kapila Shekhawat, S. S. Rathore, O. P. Premi, B. K. Kandpal, and J. S. Chauhan

Directorate of Rapeseed-Mustard Research, Sewar, Rajasthan Bharatpur 321 303, India

ABSTRACT

India is the fourth largest oilseed economy in the world. Among the seven edible oilseeds cultivated in India, rapeseed-mustard contributes 28.6% in the total oilseeds production and ranks second after groundnut sharing 27.8% in the India's oilseed economy. The mustard growing areas in India are experiencing the vast diversity in the agro climatic conditions and different species of rapeseed-mustard are grown in some or other part of the country. Under marginal resource situation, cultivation of rapeseed-mustard becomes less remunerative to the farmers. This results in a big gap between requirement and production of mustard in India. Therefore site-specific nutrient management through soil-test recommendation based should be adopted to improve upon the existing yield levels obtained at farmers field. Effective management of natural resources, integrated approach to plant-water, nutrient and pest management and extension of rapeseed-mustard cultivation to newer areas under different cropping systems will play a key role in further increasing and stabilizing the productivity and production of rapeseed-mustard. The paper reviews the advances in proper land and seedbed preparation, optimum seed and sowing, planting technique, crop geometry, plant canopy, appropriate cropping system, integrated nutrient management and so forth to meet the ever growing demand of oil in the country and to realize the goal of production of 24 million tonnes of oilseed by 2020 AD through these advanced management techniques.

INTRODUCTION

Rapeseed-mustard is the third important oilseed crop in the world after soybean (Glycine max) and palm (Elaeis guineensis Jacq.) oil. Among the seven edible oilseed cultivated in India, rapeseed-mustard (Brassica spp.) contributes 28.6% in the total production of oilseeds. In India, it is the second most important edible oilseed after groundnut sharing 27.8% in the India's oilseed economy. The share of oilseeds is 14.1% out of the total cropped area in India, rapeseed-mustard accounts for 3% of it. The global production of rapeseed-mustard and its oil is around 38–42 and 12–14 mt, respectively. India contributes 28.3% and 19.8% in world acreage and production. India produces around 6.7 mt of rapeseed-mustard next to China (11-12 mt) and EU (10–13 mt) with significant contribution in world rapeseed-mustard industry. The rapeseed-mustard group broadly includes Indian mustard, yellow sarson, brown sarson, raya, and toria crops. Indian mustard (Brassica juncea (L.) Czernj. & Cosson) is predominantly cultivated in Rajasthan, UP, Haryana, Madhya Pradesh, and Gujarat. It is also grown under some nontraditional areas of South India including Karnataka, Tamil Nadu, and Andhra Pradesh. The crop can be raised well under both irrigated and rainfed conditions. Brown sarson (B. rapa ssp sarson) has 2 ecotypes lotni and toria. Yellow sarson (B. rapa var. trilocularis) is cultivated in Assam, Bihar, Orissa, and West Bengal as rabi crop. In Punjab, Haryana, UP, Himachal Pradesh, and Madhya Pradesh, it is grown mainly as a catch crop. Taramira (Eruca sativa) is grown in the drier parts of North-West India comprising the states of Rajasthan, Haryana, and UP. Gobhi sarson (B. napus L. ssp. oleferia DC. varannua L.) and karan rai (Brassica carinata) are the new emerging oilseed crops having limited area of cultivation. Gobhi sarson is a long duration crop confined to Haryana, Punjab, and Himachal Pradesh. It has good yield potential, wide adaptability and possesses high oil content of good quality. Karan rai yields well and shows better environment adoption and substantial resistance to pests and diseases. The country witnessed yellow revolution through a phenomenal increase in production and productivity from 2.68 MT and 650 kg/ha in 1985-86 to 6.96 MT and 1022 kg/ha in 1996-1997, respectively. In spite of these achievements, there exists a gap between production potential and actual realization. In India rapeseed-mustard is grown on an area of 5.53 Mha with production and productivity of 6.41 MT and 1157 Kg/ha, respectively [1].

Mustard is cultivated in mostly under temperate climates. It is also grown in certain tropical and subtropical regions as a cold weather crop. Indian mustard is reported to tolerate annual precipitation of 500 to 4200 mm, annual temperature of 6 to 27°C, and pH of 4.3 to 8.3. Rapeseed-mustard follows C_3 pathway for carbon assimilation. Therefore, it has efficient photosynthetic response at 15–20°C temperature. At this temperature the plant achieve maximum CO_2 exchange range which declines thereafter. Rai is mostly grown as a rainfed crop, moderately tolerant to soil acidity, preferring a pH from 5.5 to 6.8, thrives in areas with hot days and cool night and can fairly sustain drought. Mustard requires well-drained sandy loam soil. Rapeseed-mustard has a low water requirement (240–400 mm) which fits well in the rainfed cropping systems. Nearly 20% area under these crops is rainfed. A review is prepared on advances on agronomic practices for enhancing the rapeseed-mustard production in India. A review of the work done on the different aspects in India and abroad especially under advance agronomic practices is done in this paper.

CROP ADAPTATION AND DISTRIBUTION

The rapeseed-mustard group includes brown sarson, raya, and toria crops. Indian mustard (Brassica juncea(L.) Czernj. & Cosson) is predominantly cultivated in Rajasthan, UP, Haryana, Madhya Pradesh, and Gujarat. It is also grown under some nontraditional areas of South India including Karnataka, Tamil Nadu, and Andhra Pradesh. The crop can be raised well under both irrigated and rainfed conditions. Being more responsive to fertilizers, it gives better return under irrigated condition. Brown sarson (B. rapa ssp. sarson) has 2 ecotypeslotni and toria. Yellow sarson (B. rapa var. trilocularis) is cultivated in Assam, Bihar, Orissa, and West Bengal as rabi crop. In Punjab, Haryana, UP, Himachal Pradesh, and Madhya Pradesh, it is grown mainly as a catch crop. Taramira (Eruca sativa) is grown in the drier parts of North-West India comprising the states of Rajasthan, Haryana and UP. Gobhi sarson (B. napus l. ssp. oleferia DC. Var. annua L.) and karan rai (Brassica carinata) are the new emerging oilseed crops having limited area of cultivation. Gobhi sarson is a long duration crop confined to Haryana, Punjab, and Himachal Pradesh. It is photo- and thermosensitive and makes little growth up to middle of February, but in the end of this month, plants make a quick growth. It has good yield potential, wide adaptability, and possesses high oil content of good quality. There are eight cultivated crops in rapeseed-mustard crop; the main characteristics features have been explained in Table 1.

Table 1: Salient features of cultivated species of rapeseed-mustard (Cruciferous) group of crops

SN	Common name	Botanical name	Days to maturity (days)	Yield potential, Kg/ha	Oil %
(1)	Indian mustard	Brassica juncea	105–160	1500–3000	38–42
(2)	Yellow mustard	Brassica rapa var. yellow sarson	120–155		41–47
(3)	Brown sarson	Brassica campestris	100–235	900–2000	40–45

		syn. B. rapa var. brown sarson			
(4)	Black mustard	Brassica nigra	70–90	1000–1200	40-41
(5)	Karan rai	Brassica carinata	150–200		36–43
(6)	Toria	Brassica rapa var. toria	70–100	600–1800	36–44
(7)	Taramira	Eruca sativa	140–150	700–1400	34–38
(8)	Gobhi sarson	Brassica napus	145–180	1300–2700	37–45

Karan rai also yields well under a wide range of climate partly because it has a large number of primary and secondary racemes. It shows better environment adoption and substantial resistance to pests and diseases. Mustard is cultivated in most temperate climates. It is also grown in certain tropical and subtropical regions as a cold weather crop. Indian mustard is reported to tolerate annual precipitation of 500 to 4200 mm, annual temperature of 6 to 27°C, and pH of 4.3 to 8.3. Rai is mostly grown as a rainfed crop, moderately tolerant to soil acidity, preferring a pH from 5.5 to 6.8, thrives in areas with hot days and cool night, and fairly resistant to drought. Mustard requires good sandy loamy soil. The agro-climatic conditions of various locations under study have been explained in Table 2.

Table 2: Agroclimatic conditions of various locations during mustard crop season

Location	Longitude	Latitude	Temp, °C		Rain fall, mm	RH %		Soil texture	Soil fertility, Kg/ha		
			Max	Min		Max	Min		N	P	K
Hisar	75°436 E	29°911 N	3.2	34.2	50–200	38	96	Sany loam	130	12	480
Pantnagar	79°2436 E	28°5812 N,	4.8	32.3	150–400	47	92	Clay loam	155	15	310
Dholi	85°3522 E	26°02.2 N	6.6	33.3	200–550	52	94	Clay loam	140	12.5	275
Ludhiana	75°18 E	30°34 N	3.5	32.0	30–120	45	95	Loamy sand	150	24	220
Bhubneshwar	85°50 E	20°16 N	14.8	34.8	180–250	38	94	Clay loam	130	19	175

VARIETALS DEVELOPMENT

Since, there is a vast variability in the climatic and edaphic conditions in the mustard growing areas of India, the selection of appropriate cultivars is important as it helps in increasing the productivity. Introduction of relatively short duration cultivar found favor with the environment where effective growing seasonal length is short. Improved varieties of mustard stabilize oil and seed yield through insulation of cultivars against major biotic and abiotic stresses enhance oil (low erucic acid) and seed meal (low glucosinolate) quality. The first Indian mustard hybrid, named "NRCHB-506," has been developed at Directorate of Rapeseed-Mustard Research, Bharatpur which can catapult the output of the country's key oil crop. The new hybrid is meant for cultivation in Rajasthan and Uttar Pradesh. Other high yielding varieties include "JM-1," "JM-3," and "Pusa Bold," "NRCDR-2," "NRCDR 601." Their yield potentials vary from 16 to 25 q/ha. At IARI, an early-maturing and bold seeded mustard variety has been developed called "Mehak" (B. juncea). This improved variety is suitable for early sowing to replace toria (B. rapa var. toria) in Delhi and adjoining areas. Gobhi sarson has a good yield potential, wide adaptability and possesses high oil content of good quality. "Hyola" (PAC-401) is canola type hybrid rapeseed, developed in India by Advanta India Ltd, Holland-based multinational company. "Neelam" (HPN-3) and "Sheetal" (HPN-1) are the popular varieties of gobhi sarson [2]. Since inception of mustard research programme in India, number of tolerant varieties to various abiotic and biotic stresses of rapeseed-mustard has been developed (Table 3).

Table 3: Varieties tolerant to various abiotic and biotic stresses of mustard (Brassica juncea)

SN	Specific abiotic/ biotic stress	Tolerant verities
(1)	Rainfed	Aravali, Geeta, GM 1, PBR 97, PusaBahar, Pusa Bold, RH 781, RH 819, RGN 48, Shivani, TM 2, TM 4, Vaibhav, RB 50
(2)	Salinity tolerant	CS 52, CS 54, Narendra Rai (NDR8501)
(3)	Frost tolerant	RGN 13, RH 819, Swaranjyoti, RH 781, RGN 48

(4)	High temperature tolerant	Kanti, Pusa Agrani, RGN 13, Urvashi, NRCDR 02, Pusa mustard 25 (NPJ 112), Pusa mustard 27 (EJ 17)
(5)	White rust resistant	Basanti, JM 1, JM 2, NRCDR-2
(6)	Alternaria blight tolerant	Jawahar Mustard 3, Him Sarson 1 (ONK 1), Ashirwad (RK-01-03)

"Pusa Jaikisan" of B. juncea is the first variety though tissue culture. "TL-15," a toria variety has been recommended as summer crop for high altitude of Himachal Pradesh. In an attempt to incorporate resistance/tolerance to biotic and abiotic stresses in high yielding varieties, aphid tolerant strains like "RH-7846," "RH-7847," "RH-9020" and "RWAR-842," Alternaria blight moderately resistant variety "Saurabh"; white rust resistant variety, "Jawahar Mustard-1"; salt tolerant varieties "Narendra Rai" and "CS-52" frost tolerant "RH-781" and "RH-7361" varieties have been identified. "RH-781" is also drought tolerant and suitable for intercropping. For nontraditional areas, Indian mustard varieties "Rajat," "Pusa Jaikisan" and "Sej.2" have been recommended.

LAND AND SEEDBED PREPARATION

A mustard seedbed should be firm, moist, and uniform which allows good seed-to-soil contact, even planting depth and quick moisture absorption leading to a uniform germination. Tillage affects both crop growth and grain yield. The various tillage systems are as follows: conventional tillage includes moldboard ploughing followed by disc harrowing; reduced tillage includes disc ploughing followed by disc harrowing and complete zero tillage in which crop is sown under uncultivated soil. Minimum tillage, with or without straw, enhances soil moisture conservation and moisture availability during crop growth. As a consequence, the root mass, yield components and seed yield increase [3]. Zero tillage is preferred in mustard as it conserves more moisture in the soil profile during early growth period. Subsequent release of conserved soil moisture regulates proper plant water status, soil temperature, lower soil mechanical resistance, leading to better root growth and higher grain yield of mustard [4]. Success with minimum or zero tillage requires even distribution of crop residues, as a well-designed crop rotation and evenly distributing residue will create a firm, moist and uniform seedbed.

Continuous zero tillage results in redistribution of extractable soil nutrients with greater concentration near the soil surface, compared with conventional tillage where mixing of soil, residues, fertilizers, and lime results in a relatively homogeneous soil to the depth of tillage [6]. With zero tillage having greater root density in the surface soil but lesser root density below a depth of 15 cm in the soil profile. Therefore, P and K uptake by crops grown under zero tillage is greater than those grown by conventional methods. But the plant growth and dry matter yields of mustard under zero tillage will be higher only if N fertilizers are applied in appropriate amount [7]. Under AICRP on RM at Dholi, Kanke, Bhubaneshwar, and Behrampur maximum seed yield oftoria and mustard was obtained in line sowing under zero tillage practice which indicated that mustard can be grown well under zero tillage.

At Bhubaneshwar, line sowing of mustard under zero tillage after rice gave the maximum seed yield (933 kg/ha) and oil content (38.4%) (Table 4). The soil under zero tillage system contains higher amount of organic matter having more carbohydrate, amino acid and amino sugar that results in qualitative and quantitative improvement in soil and soil structure due to least soil disturbance. Energy output and input ratio are higher in zero tillage as compared to conventional tillage.

Table 4: Seed yield (kg/ha) and oil content (%) of toria as influenced by different N levels in utera cropping system at Bhubaneshwar

Cropping system	N levels (kg/ha)		
	0	40	80
Rice: yellow sarson (broadcast) in utera cropping (at dough stage of rice)	428 (33.3)	823 (40.3)	810 (37.6)
Rice: yellow sarson (broadcast) in utera cropping (sowing before harvest of rice)	530 (30.2)	729 (38.2)	642 (37.1)
Rice: yellow sarson (line sowing) under zero tillage in rice field	506 (34.4)	924 (41.5)	886 (39.6)
Rice: yellow sarson (line sowing) after land preparation in rice fields	388 (32.5)	846 (40.4)	820 (38.4)
Rice: yellow sarson (broadcast) after land preparation in rice fields	301 (28.2)	460 (37.6)	440 (35.5)

CD at 5% cropping system: 79 (0.7), N levels: 32 (0.4), Cropping system × N levels: 98 (1.0). Figures in the parenthesis denote oil content (%).

Source: AICRP-RM, 2003 [5].

AND SOWING

Vigorous seedling growth, good root development, early stem elongation, rapid ground covering ability, and early flowering and radiation are important yield determining traits under low temperature and radiation regime. These traits can be successfully exploited in mustard if a good seed is grown at appropriate time along with maintaining an optimum plant population.

Seed Priming

Seed treatment is a useful practice for healthy plant growth. Seed priming through controlled hydration and dehydration enhances early germination of mustard seed in less time, even in compacted soil [8]. The soaking of mustard seeds in 0.025% aqueous pyridoxine hydrochloride solution for 4 hours improved germination. The combination of pyridoxine + $N_{60}P_{20}$ + $N_{15}P_5$ (top dressing) accelerated the crop performance by enhancing seed yield and oil yield by 15.8 and 13.5%, respectively, over the control [9]. The differential response of varieties for imbibition gives advantage to some of them to germinate early as compared to others. At Hisar, maximum rate of imbibition was reported in "NRCDR-2" (41.7%) and minimum in "NRCDR-509" (7.5%). Such drastic difference in rate of imbibition is important for identification of suitable varieties under abiotic stress conditions namely drought, frost, and temperature abnormalities.

Sowing Time

Sowing time is the most vital nonmonetary input to achieve target yields in mustard. Production efficiency of different genotypes greatly differs under different planting dates. Soil temperature and moisture influence the sowing time of rapeseed-mustard in various zones of the country. Sowing time influences phenological development of crop plants through temperature and heat unit. Sowing at optimum time gives higher yields due to suitable environment that prevails at all the growth stages. Though different varieties have a differential response to date of sowing, mustard sown on 14 and 21 October took significantly more days to 50% flowering (55 and 57) and maturity (154 and 156) compared to October 7 planting [10]. Delayed sowing resulted in poor growth, low yield, and oil content. The reduction in yield was maximum in "RH-30" and minimum in "Rajat" [11, 12].

Date of sowing influence the incidence of insect-pest and disease also. Sowing on October 21 resulted in leastSclerotinia incidence [13]. The maximum (20.5–25.4°C) and minimum (3.9–10.7°C) temperatures at the flowering stage of crops established through sowing on October 21 were negatively correlated with the development of Sclerotina stem rot. Mustard aphid (Lipaphis erysimi (Kaltenbach)) has been reported as one of the most devastating pests in realizing the potential productivity of Indian mustard. Normal sowing (1st week of November) also helps in reducing the risk of mustard aphid incidence.

Planting Technique

Sowing technique depends upon land resources, soil condition, and level of management and thus broadcast, line sowing, ridge and furrow method and broad bed and furrow method are common sowing techniques. At higher soil moisture regimes, broadcasting followed by light planking gives early emergence and growth. Under normal and conserved moisture regime, seed placement in moist horizon under line sowing becomes beneficial.

At Shillongani, broadcast method was found to be more successful. Significantly higher seed yield of toria (Brassica rapa var. toria) was harvested in broadcast sowing of toria over other practices. Toria broadcast at dough stage along with 80 kg N/ha gave the highest yield (AICRP-RM, 2006). At Bhubaneshwar, line sowing of yellow sarson after land preparation produced maximum seed yield (870 kg/ha) with 40 kg N/ha [14]. At Behrampore, 40% higher seed yield of toria was obtained when sown in line after land preparation in the rice-based cropping system over broadcast (AICRP on RM, 2006). Paira or utera is a method of cropping in which the sowing of next crop is done in the standing previous crop without any tillage operation. Mustard sowing under paira/utera in the rice field has shown its edge over line sowing and broadcasting (Sowing of seeds by broad casting the seeds in the field) in eastern parts of India. At Dholi, mustard sown with paira cropping recorded significantly higher seed yield (1212 kg/ha) over line sown and broadcast method, while these 2 methods yielded at par. At Bhubaneswar, significantly higher yield (887 kg/ha) of mustard was recorded when sown as utera crop over line and broadcast sown crop [15].

Ridge and furrow sowing was superior to conventional flat sowing for growth parameters and yield of Brassica juncea [16]. Under saline condition, seed yield of canola in ridge sowing was higher by 45, 31, and 28% than broadcast, drill and furrow sowing methods, respectively [17]. The highest yield was associated with less saline environment at the ridges which allowed the seed to germinate and increase the yield. Transplanting of mustard has also been reported thereby saving time, and resources. Transplanting reduces days to maturity and results in higher seed yield. Ridge transplanting reduced water applied by 30% for each furrow as compared to 45 cm row spacing in flat method without any loss in seed yield. The corresponding increase in water use efficiency (WUE) was 27%. In bed planting, there was a 35% saving in water resulting in 32% increase in WUE (Figure 1).

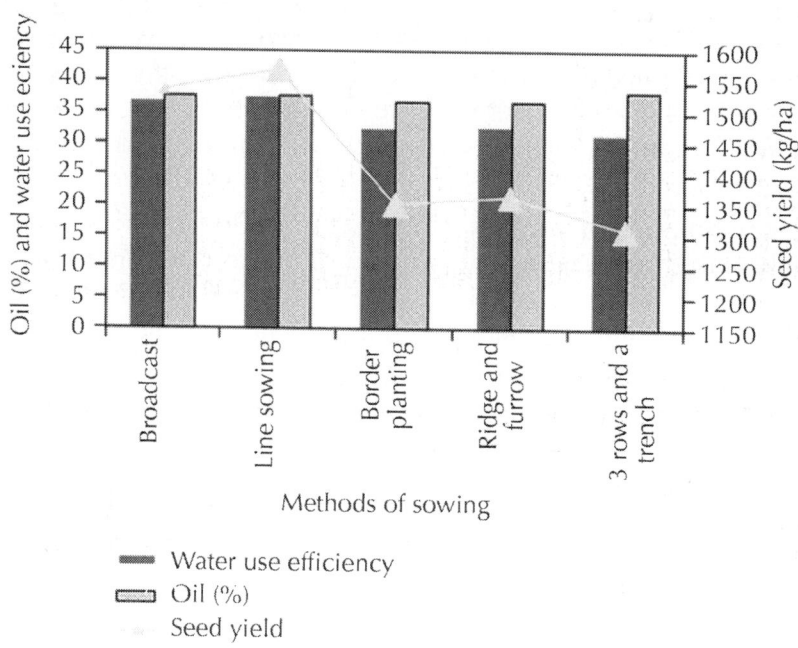

Figure 1: Seed yield, water use efficiency, Kg/ha-mm (WUE), and oil content of mustard (Brassica juncea) as influenced by various planting methods.

Crop Geometry

The competitive ability of a rapeseed-mustard plant depends greatly upon the density of plants per unit area and soil fertility status. The optimum plant population density/unit area varies with the environment, the genotype, the seeding time, and the season. Uniform distribution of crop plants over an area results in efficient use of nutrients, moisture, and suppression of weeds leading to high yield. In wider row spacing, solar radiation falling within the rows gets wasted particularly during the early stages of crop growth whereas in closer row spacing upper part of the crop canopy may be well above the light saturation capacity but the lower leaves remain starved of light and contribute negatively towards yield.

Gobhi sarson (Brassica napus) being more vigourous, the days to maturity, plant height, branches, pod, seed weight per plant, seed index, seed yield, and oil content were higher at 60 cm row spacing [18]. An increase in rows up to 30 cm correspondingly prolonged maturity days followed by optimum 45 cm and wider rows 60 cm spacing. The plants receiving narrow row spacing increased vegetative growth. Due to shade and competition for nutrients and moisture the crop matures later by increasing developmental phases Taller plants were observed in the plots where crop was planted in rows of 60 cm apart followed by 45 cm and 30 cm row spacing due to sufficient space resulting in plants grown well and showed greater height [19] (Gupta, 1988). The regression coefficient indicated that each increase in row spacing up to 60 cm resulted in increased crop maturity by 0.54 days, plant height by 0.44 cm, number of branches would increase by 0.11, pods per plant by 1.96, seeds per pod by 0.04, seed weight per plant by 0.45, seed index by 0.152 g, oil content by 0.8% and increase in seed yield by 10.32 kg/ha. The recommended spacing for mustard is 30 × 10 and for hybrids it is 45 × 10. At Kumher, plant spacing 45 × 15 recorded significant higher seed yield over other spacing but was on a par with 45 × 10 cm. At Pantnagar, 30 × 15 recorded significantly higher seed yield which remained on a par with 45 × 10 and 45 × 15 cm plant spacing [20].

Plant Population and Inter-Plant Shading

The dense plant population reduces the yield due to reduction in the photosynthetically active leaf area caused by mutual shading. In an experiment on Brassica juncea (Var. laxmi) the reduction is more due to shading at 91–110 DAS over 71–90 DAS. The specific leaf weight (SLW), crop growth rate (CGR), and net assimilation rate (NAR) were more adversely affected by 50% shading at 71–90 DAS. Net assimilation ratio remained unaffected by 25% shading, while it reduced significantly by 50% shading at both the stages; the reduction was more with 50% due to shading at 91–110 DAS. On an average 50% shading at 91–110 DAS was more deleterious than 25% shading at 91–110 DAS, that is, at terminal seed development stage (Table 5).

Table 5: Effect of shading on yield and growth parameters in Indian mustard at Hisar

Treatment	Seed yield (kg/ha)	SLW (mg/cm2)	CGR (g/m2/day)	NAR (mg/m2/day)
Control	571.6	8.3	11.3	0.93
25% shading at 71–90 DAS	546.0	7.0	10.8	0.93
25% shading at 91–110 DAS	490.9	7.9	9.4	0.87
50% shading at 71–90 DAS	527.0	6.4	9.9	0.95
50% shading at 91–110 DAS	380.0	7.0	8.3	0.83
CD at 5%	33.1	1.3	1.0	0.05

Source: AICRP-RM, 2004.

CROPPING SYSTEM

Physiography, soils, geological formation, climate, cropping pattern, and development of irrigation and mineral resources greatly influence selection of variety and cropping system. Fallow mustard is popular sequence in major mustard growing areas but studies show that some of the crop result in better resource utilization and high remuneration if included in mustard-based cropping system.

Mustard Productivity under Various Crop Sequences

Under AICRP trials at Dholi, fallow-mustard sequence gave significantly higher seed yield which was on a par with blackgram-mustard sequence: urdbean-mustard at Morena; greengram-mustard, guar-mustard, and pearl-millet-mustard at S. K. Nagar and Hisar; maize-mustard at Kangra and Pantnagar revealed superiority to fallow-mustard. The productivity of the system also depends upon the fertility status and the nutrient supply. When mustard was grown after soybean or bajra, the response to S was observed up to 40 kg S/ha [21]. Productivity measured in terms of land equivalent ratio (LER) was higher for intercropping of chickpea and mustard in the 4:1 row ratio than for sowing of chickpea and mustard in sole stands [22].

Inclusion of Gobhi Sarson (Brassica Napus) Under Various Cropping Sequences

Gobhi sarson is comparatively recent introduction and hence needs identification of suitable cropping systems. Growing gobhi sarson and toria in alternate rows at 22.5 cm spacing is very remunerative. Maize-gobhi sarson, blackgram-gobhi sarson, rice-gobhi sarson, and soybean-gobhi sarson were identified remunerative cropping systems at Kangra [21].

Mustard-(Brassica Juncea) Based Cropping System under Rainfed Areas

There are possibilities of increasing cropping intensity in monocropping mustard areas under rainfed condition. Green manuring or guar during rainy season enhance seed yield of succeeding mustard [12]. In addition to efficient resource use, intercropping imparts stability to productivity and reduces the risk of crop failure. Under irrigated conditions, at Bharatpur, the seed yield equivalent of mustard (Brassica juncea) was significantly higher where mustard was grown in combination with potato (1 : 3), mustard + wheat (1 : 5), mustard + barley (1 : 5) than pure mustard. At Hisar, intercropping Brassica juncea (variety RH-30) with rabi crops had revealed highest gross return (Rs. 29,498) when mustard was grown as a pure crop. The mustard seed equivalent was highest in mustard + chickpea (1 : 5). Intercropping of mustard with chickpea, field pea, or linseed proved superior over their cultivation as a pure crop (Table 6).

Table 6: Seed yield (kg/ha), mustard equivalent yield (MEY), and gross return (Rs./ha) as influenced by various intercropping combinations under rainfed conditions at Hisar

Treatment	Main crop	Intercrop	MEY	Gross return (Rs./ha)
Pure mustard	2565	—	2565	29,497
Mustard + chickpea (1 : 5)	966	1035	1956	22,494
Mustard + fieldpea (1 : 5)	1002	189	1230	14,145
Mustard + linseed (1 : 5)	996	642	1721	19,791
Mustard + lentil (1 : 5)	1015	—	1015	11,672
Mustard without intercropping at same distance as in intercropping	1097	—	1092	12,668
CD at 5%	—	—	350	—

Source: AICRP-RM, 1997 [11].

FERTILIZER MANAGEMENT

Adequate nutrient supply increases the seed and oil yields by improving the setting pattern of siliquae on branches, number of siliquae/plant, and other yield attributes [23]. Recommended dose of fertilizers (RDF) for different zones changes with climate, soil type, time, and type of cropping system followed.

Nitrogen and Phosphorus Fertilization

Nitrogen use efficiency is greatly influenced by the rate, source, and method of fertilizer application. The rate of nitrogen depends upon the initial soil status, climate, topography, cropping system in practice, and crop. Crop under zero tillage is also more productive (695 kg/ha) with 80 kg N/ha [14]. Increase in the nitrogen level up to 60 kg N/ha consistently and significantly increased the number of primary branches, number of seeds per siliquae and 1000 seed weight [24]; however, increasing the nitrogen level up to 90 kg/ha increased the number of secondary branches per plant, number of siliquae per plant, and seed and straw yield with maximum cost benefit ratio of 3.03 [25]. Split application of total nitrogen in three equal doses one-each as basal, second after first irrigation and remaining one-third after second irrigation resulted in maximum increase in yield attributes and yield of Brassica juncea compared to application of total nitrogen in two split doses [26]. Top dressing of N fertilizers should be done immediately after first irrigation. Delaying of first irrigation, results in yield reduction of mustard crop. The application of nitrogen with presowing irrigation was superior to that of nitrogen application with last preparatory tillage. In case of nitrogen applied with pre-sowing irrigation single application of nitrogen was on a par with split application [27].

Application of phosphorus up to 60 kg/ha significantly enhanced dry matter/plant. Plant height, branches per plant and leaf chlorophyll content increased with up to 40 kg P/ha. The uptake of NPK and sulphur by both seed and stover increased significantly with successive increase in nitrogen levels up to 120 kg N/ha, sulphur levels up to 60 kg S/ha, and P_2O_5 level up to 60 kg P_2O_5/ha. Seed yield and yield attributes increased while oil content decreased with increasing level of nitrogen up to 120 kg/ha. Different levels of phosphorus increased seed yield, maximum being at 80 kg P/ha due to higher number of secondary branches/plant and consequently siliquae/plant. Oil content also increased with increase in levels of N, P_2O_5, and S. Activities of all nitrogen assimilating enzymes, namely; nitrate reductase, nitrite reductase, glutamine synthetase, and glutamate synthetase were found to be maximum at 100 kg N/ha.

Sulphur Fertilization

Among the oilseed crops, rapeseed-mustard has the highest requirement of sulphur [28]. Sulphur promotes oil synthesis. It is an important constituent of seed protein, amino acid, enzymes, glucosinolate and is needed for chlorophyll formation [29]. Sulphur increased the yield of mustard by 12 to 48% under irrigation, and by 17 to 124% under rainfed conditions [30]. In terms of agronomic efficiency, each kilogram of sulphur increases the yield of mustard by 7.7 kg [31].

Oil content in Canola-4 and Hyola-401 is 3% higher than the hybrid "PGSH-51" due to the effect of various doses of nitrogen and sulphur, while the oleic acid content in these hybrids is double that "PGSH-51." "PGSH-51" had erucic acid ranging from 23.2 in to 29.4%. At higher sulphur level there is 2-3% reduction in erucic acid content. However, lower level of nitrogen reduced erucic acid content by 3% with a concomitant increase in oleic acid (Table 7). Higher doses of sulphur along with low doses of nitrogen affect the chain elongation enzyme system thereby leading to reduction in erucic synthesis.

Table 7: Effect of N and S levels (kg/ha) application on fatty acid composition and glucosinolate content in Brassica juncea cv. Varuna at Ludhiana

N (Kg/ha)	S (Kg/ha)	Glucosinolate content (µmoles/g in defatted meal)	Palmitic acid	Stearic acid	Oleic acid	Linoleic acid	Linolenic acid	Eicosenoic acid	Erucic acid
75	0	64	2.61	1.17	11.78	14.99	6.48	50.91	11.80
75	20	72	2.88	1.31	10.15	14.53	5.14	52.75	12.28
100	0	52	2.58	1.58	13.16	15.31	7.01	49.55	10.57
100	20	42	2.91	1.65	11.94	15.06	6.13	49.63	12.18
125	0	52	3.01	1.33	12.19	16.17	5.91	47.71	12.26
125	20	42	4.42	1.31	16.12	16.55	6.57	44.77	9.55

Source: AICRP-RM, 2007 [14].

A significant increase in yield was observed with increase in sulphur levels up to 40 kg S/ha in mustard-based cropping system. At Bawal, the highest seed yield of mustard was recorded in green gram-mustard cropping sequence while the lowest (2686 kg/ha) in pearl millet-mustard sequence. In rice-mustard sequence, the optimum seed yield of mustard was obtained at 40 kg S/ha at Behrampore and for blackgram-mustard at Dholi. Each successive increase in S level increased seed yield up to 20 kg S/ha at Dholi and Ludhiana, 40 kg S/ha at S. K. Nagar, and 60 kg S/ha at Behrampore and Morena conditions [32].

Micronutrients

Mustard, in general is very sensitive to micronutrient deficiency, specially zinc and boron. The increase in seed yield was 8.5% at 12.5 kg $ZnSO_4$/ha. The harvest index (HI) was significantly affected by Zn application, although seed yield showed diminishing return with additional $ZnSO_4$ doses (Table 8).

Table 8: Effect of Zn on yield and yield attributes of indian mustard

ZnSO4 (Kg/ha) levels	Seed yield (kg/ ha)	Secondary branches/ plant	Oil content (%)	Oil yield (kg/ha)	Protein (%)	Protein yield (kg/ha)	Harvest index (%)
0	1161	6.5	40.2	465.6	22.1	255.2	21.6
12.5	1260	8.1	39.9	501.1	22.5	281.9	22.4
25.0	1336	9.6	39.9	532.4	22.6	301.6	22.9
50	1414	12.4	39.9	570.0	22.5	318.6	22.2
CD at 5%	33	0.7	NS	22.8	NS	18.8	0.8

Source: AICRP-RM, 2000 [33].

The response of various ideotype to the applied micronutrients varies considerably. The response of Indian mustard varieties, viz. 'Pusa Bold' and 'Vardan' to applied zinc was found higher (AICRP-RM, 2000) as compared to Varuna, RH- 30 and Aravali (Figure 2).

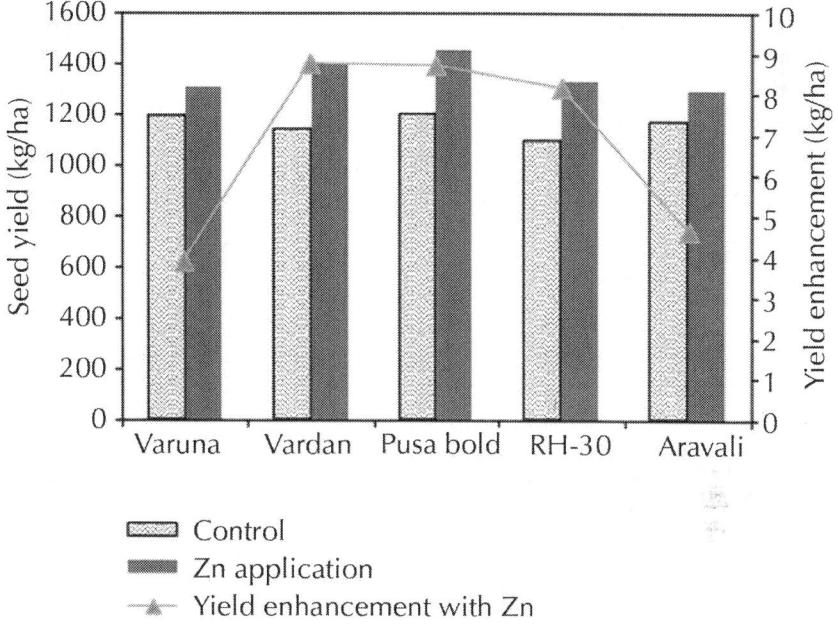

Figure 2: Influence of zinc application on seed yield of different cultivars of mustard.

The concentration of Zn at flowering, pod formation stage, concentration and uptake of Zn in straw and grain at maturity and uptake of Zn in grain and straw at maturity of Indian mustard increased significantly with increase in Zn levels [34]. Similarly, the seed yield increased significantly (16–47%) with the application of boron. The average response to boron application ranged from 21 to 31%. The yield increase was due to 27% and 10% increase, respectively, in seeds/siliqua and 1000 seed weight, indicating the importance role in seed formation [35, 36].

Organic Sources of Nutrients

Bulky organic manures are applied to improve overall soil health and reduce evaporation losses of soil moisture. Depending upon the availability of raw material and land use conditions various organic sources,namely, clusterbean (green manure), Sesbania (green manure), mustard straw @ 3 t/ha and Vermicompost (2.5–7.5 t/ha) have been evaluated at Bharatpur. Green manure with Sesbania gave significantly higher mustard seed yield at Bharatpur and Bawal. Sesbania green manuring has shown higher mustard yield and improved soil environment (AICRP-RM, 2006).

Many biostimulants also encourage higher production. At Hisar, foliar spray of Bioforce (an organic formulation) 2 mL/L at the flowering and siliqua formation stage enhanced mustard seed yield (2059 kg/ha) [14].

Integrated Nutrient Management (INM)

It is important to exploit the potential of organic manures, composts, crop residues, agricultural wastes, biofertilizers and their synergistic effect with chemical fertilizers for increasing balanced nutrient supply and their use efficiency for increasing productivity, sustainability of agriculture, and improving soil health and environmental safety. Balanced fertilization at right time by proper method increases nutrient use efficiency in mustard. Experiments have been conducted at different AICRP centres with the integrated use of organic manure, green manure, crop residue, and biofertilizers along with inorganic fertilizers. INM not only reduces the demand of inorganic fertilizers but also increases the efficiency of applied nutrients due to their favourable effect on physical, chemical and biological properties of soil. The introduction of leguminous crops in the rotational and intercropping sequence and use of bacterial and algal cultures play an important role in increasing the nutrient use efficiency [37].

Growth Promoter, BioFertiliser as a Component of INM

Biofertilizers are inoculants or preparation containing micro-organims that apply nutrients especially N and P. Two types of N-fixing microorganisms namely free living (Azotobacter) and associative symbiosis (Azospirillum) and two P supplying microorganisms, namely, phosphate solubilizing bacteria and vesicular arbuscular mycorrhiza (VAM) were extensively tested at various AICRP-RM centers. Inoculation of mustard seeds with efficient strains of

Azotobacter and Azospirillum enhanced the seed yield up to 389 and 305 kg respectively with 40 Kg N/ha. The total NPK uptake was also higher with Azotobacter inoculation. The combined application of 10 t FYM + 90:45:45 NPK kg/ha with Azotobacter inoculation gave the highest B:C ratio of 1.51. At lower N levels, without inoculation, the seed yield decline was more as compared to inoculated treatment. Growth promoter's formulations like bioforce and biopower contain bio-amino acid, plant growth promoting terpenoid, siderophores, and attenuated bacteria fortified with BGA helped to increase water and nutrient absorption from the soil. Similarly, bioforce contains natural free amino acid, phytohormones, macro- and microelements and plant growth promoting terpenoid activated the cell division and stimulates plant growth, development, and photosynthate translocation. RDF (80:40:0) along with 25 kg Biopower/ha + spray of Bioforce (1 l in 500 litres of water) at 50% flowering and pod filling stage gave significant higher yield of mustard over other combinations [35, 36].

Effect of INM on Quality of Mustard Oil

At Kanpur, INM studies were evaluated in maize-mustard, bajra-mustard, and fallow mustard sequence. In maize-mustard sequence, 100/75% of RDF + 2 t FYM gave highest seed yield and quality of the oil (Table 9).

Table 9: Effect of INM on quality of mustard (Kanti-RK 9807) under maize-mustard sequence

Treatment	Legends	Oil content (%)	Fatty Acid composition (%)						
			16:1 Palmtic acid	18:1 Oleic acid	18:2 Linoleic acid	18:3 Linolenic acid	20:1 Eicosenoic acid	22:1 Erucic acid
RDF (120-40-40)	T1	40.4	2.8	18.4	10.1	10.6	4.3	52.7
T1 + 10t FYM/ha	T2	40.9	2.8	16.3	13.3	10.4	4.1	52.2
T2 + 40Kg S/ha	T3	40.4	2.9	18.0	14.4	12.2	3.2	48.6
T3 + Zn SO4 25kg/ ha	T4	40.3	2.8	17.8	14.9	10.1	6.1	47.3
T4 + B 1 kg/ha	T5	40.7	2.7	23.0	16.2	9.0	5.2	43.3
T1 + Crop residue (Maize)	T6	40.1	2.7	20.0	14.3	9.2	4.4	48.6
75% RDF		40.4	2.6	17.8	15.1	7.9	6.3	49.7

Source: Modified from AICRP-RM, 2002 [21].

Integrated Nutrient Management (INM) and Nutrient Use Efficiency

INM improves the nutrient uptake by mustard and hence enhances the use efficiency of various nutrients from the soil. The incorporation of 25% nitrogen through FYM + 75% by chemical fertilizer + 100% sulphur significantly enhanced the uptake use efficiency and of nitrogen and sulphur in both seed and stover of crop followed by 100% NS and 50% N through FYM + 50% by chemical fertilizer + 100% S [38]. The highest mustard-equivalent yield, which includes converted yield of other crops in to mustard seed yield based on market price of the crops (24.88 q/ha), net monetary returns (Rs. 15,537/ha), B:C ratio (2.07), and agronomic efficiency (16.1) were recorded with the application of 100% recommended N in the rainy season through FYM and 100% recommended NP in the winter season through inorganic fertilizers [39]. Agronomic efficiency is the response in terms of increase in mustard seed yield per unit use of nitrogen.

At Bharatpur and Jobner, 17.8 and 8.6% increase in seed yield was recorded with 50% RDF + 50% N through FYM and vermin-compost. Sole organic treated plot recorded 29.9% lesser seed yield over RDF at Jobner [32]. Amount of available phosphorus increased over initial value when organic manures and crop residues were incorporated. Organic carbon status builds up in organic source incorporated plots. The application of 10t FYM/ha in addition to recommended dose of fertilizer (RDF) improved soil physical condition by improving aggregation, increased saturated hydraulic conductivity, and reducing bulk density and penetration resistance of the surface soil [40].

WATER MANAGEMENT

Rapeseed-mustard crop is sensitive to water shortage. A substantial rapeseed-mustard area in Rajasthan (82.3%), Gujrat (98%), Haryana (75.6%), and Punjab (92.4%) is covered under irrigation. A positive effect of irrigating rapeseed-mustard at critical stages is observed. Water use efficiency was highest when irrigation was applied at 0.8 IW:CPE ratio and increased with increasing N rate [41, 42]. Number of irrigations is important for working out the most efficient water use by mustard. For mustard, two irrigations, one at flowering stage and at siliqua formation stage increased seed yield by 28% over the rainfed plots [43]. Increase in the amount of water increased leaf water potential, stomatal conductance, light absorption, leaf area index, seed yield, and evapotranspiration and decreased canopy temperature [44]. In similar study by Panda et al. [45], an average increase in seed yield with irrigation at the flowering and pod development stages and irrigation at the flowering stage over the control was 62.9% and 41.7%, respectively. However, for number of seeds per siliqua and oil content, single irrigation at 45 DAS remained parallel with two irrigations [46]. The water use efficiency was highest with one irrigation at 45 DAS. Crop receiving two irrigations at preflowering and pod-filling stages produce about 33 percent more seed than unirrigated crops [47]. Single irrigation given at vegetative stage is found to be most critical, as irrigation at this stage produces the highest yield. When two irrigations are given, the irrigation at vegetative and pod formation stages is of maximum benefit. The irrigation at vegetative, flowering, and pod formation stages resulted in the highest yield, where three irrigations were given. Oil and protein yield were also significantly affected by number and stages of irrigation (Table 10).

Table 10: Influence of irrigation levels and stages on seed yield, oil yield and protein yield of Indian mustard

Treatment	Seed yield (Kg/ha)	Oil yield (kg/ha)	Protein yield (kg/ha)
4 irrigations at V + F + P + S	2260	909	454
3 irrigations at V + F + P	2250	901	454
3 irrigations at V + F + S	2200	886	442
2 irrigations at V + P	2150	879	436
2 irrigations at V + F	2090	841	422
2 irrigations at F + P	2020	803	417
2 irrigations at P + S	1520	574	316
1 irrigation at V	1920	773	386
1 irrigation at F	1790	727	371
CD at 5%	480	144	94

Note: V: vegetative stage; F: flowering stage; P: pod formation; S: seed development.
Source: AICRP-RM, 1999 [15].

Irrigation is very important for getting the optimum productivity potential of mustard, but equally important is the quality of irrigation water. If the quality of irrigation water is poor, it needs certain treatment and management before being utilized for crop production. The increasing levels of salinity of the irrigation water applied at presowing and flower initiation reduces the plant height, the branching pattern, and the pod formation [48]. Irrigation with saline water (12 and 16 dS/m) decreased the dry matter yield significantly when applied at pre-sowing or later. The saline irrigation at the pre-flowering stage or later reduced the grain yield by 50% and 70%, respectively.

As a result of saline water irrigation, the soil water infiltration was reduced up to 7%. The EC and exchangeable sodium percentage (ESP) were increased by 2.2 dSm^{-1} and 9.0, respectively. The yield of mustard crop could be further increased by better leveling the plots, reducing the level difference to less than 10 cm [49]. The ill effects of saline water can be overcome with proper N management. Nonsaline water can be substituted by applying N and saline water [50].

WEED MANAGEMENT

Weeds cause alarming decline in crop production ranging from 15–30% to a total failure in rapeseed-mustard yield. The critical period is 15–40 days. Weeds compete with crop plants for water, light, space, and nutrients. Therefore, timely and appropriate weed control greatly increases the crop yield and thus nutrient use efficiency. The common weeds of mustard are Chenopodium album, C. murale, Cyperus rotundas, Cynodon dactylon, Melilotus alba, Asphodelus tenuifolius, Orobanche spp. and Anagallis arvensis.

Farmers have adopted herbicides for weed control because the chemicals can increase the profit, weed control efficiency, production flexibility and reduce time and labour requirement for weed management. Hand weeding at 20DAS, fluchloralin preplant incorporation @ 0.75 kg/ha, wooden hand plough between the lines at 35 DAS on Indian mustard was found effective [51]. Polythene mulch was also found effective in controlling the weeds in mustard [52]. At Bawal, reductions in weed population and dry matter were obtained with fluchloralin supplemented with hand weeding at 30 and 60 DAS, which remained on a par with isoproturon and pendimethalin supplemented with hand weeding at 30 and 60 DAS. Weed-free plot recorded 39.9% higher seed yield over weedy check [32].

Broomrape (Orobanche) is a major devastating parasitic weed of mustard. Broomrape weed infestation caused 28.2% average reduction in Indian mustard yield. Among Orobanche spp., O. aegyptiaca is one of the most important parasitic weed causing severe yield and quality reducing factor in rapeseed-mustard. It is endemic in semiarid region and may reach epidemic proportions depending upon soil moisture and temperature. Preceding crop of cowpea, black gram, moth bean, sunn hemp, cluster bean, and sesame significantly reducedOrobanche menace in succeeding mustard crop while sorghum, pearl millet, chilies, and green gram did not influence broomrape infestation in mustard [53]. At Bharatpur, S. K. Nagar and Bawal directed spray of glyphoste (0.25–1.0%) and 2 drops of soybean oil per young shoot of Orobanche showed effective control and recorded 91.9% higher seed yield over infected sick plot.

Some cultural practices like mulching and hoeing are also helpful to curb some of the major weeds in mustard by providing a shield against sunlight, reducing the soil temperature and acting as a physical barrier for emergence of weeds. Maximum seed yield (2540 kg/ha) was obtained in the treatments where plots were kept weed-free followed by the treatment where mulching was done after hoeing (Table 11).

Table 11: Seed yield (kg/ha) and weed population/m^2 as influenced by different weed control practices

Treatment	Seed yield	Weed population/m2
Control	1620	57.0
Weed free (Khurpi)	2520	0.0
Hoeing at 25 DAS	2300	19.3
Mulching with bajra florets	1960	23.0
Fluchloralin @ 1 kg a.i./ha PPI	2000	23.0
Pendimethalin @ 1 kg a.i./ha PE	2050	22.1
Isoproturon @ 1 kg a.i./ha PE	1740	26.3
Hoeing at 25 DAS + mulching	2400	17.9
Fluchloralin @ 1 kg a.i./ha PPI + Hoeing	2210	20.3

Fluchloralin @ 1 kg a.i./ha PPI + Mulching	2100	22.5
Pendimethalin @ 1 kg a.i./ha PE + hoeing	2300	18.9
Pendimethalin @ 1 kg a.i./ha PE + mulching	1860	19.5
Isoproturon @ 1 kg a.i./ha PE + hoeing	1950	22.5
Isoproturon @ 1 kg a.i./ha PE + mulching	1910	22.9

Source: AICRP-RM, 2002 [21].

RESPONSE TO PLANT GROWTH REGULATORS

Plant growth regulators (PGR) involved in manipulating plant developments, enhancing yield and quality have been actualized in recent years. Indeterminate plant growth habit, shattering, or dehiscence of fruits and lodging are the most significant and consistent limitations to maximum seed yields in Brassica spp. Considerable seed loss takes place, before or during harvest, due to shattering of fruits, which is correlated with hormonal imbalances and poorly developed lignified cells in the fruit wall. Further, lodging of the crop canopy adversely affects seed quality and yield due to decreased photosynthesis, increased disease severity, impaired rate of drying, and reduced harvest efficiency. Chemical plant growth regulators are being increasingly used as an aid to yield enhancement [54].

Brassinolide is the most bioactive form of the growth-promoting plant steroid termed as Brassinosteroids. Biologically active brassinosteroids show high growth-promoting as well as antistress activity besides other multiple effects on growth and development. As botanical juvenile hormones, they enhance the growth of young plant tissue and stimulate in submicromolar concentrations metabolic, differentiation and growth processes. Brassinosteroid caused accumulation of maximum total dry matter as compared to rest of the treatment at physiological maturity.

NPK accumulation and yield were maximal when spraying of GA_3 was done at 40 DAS [55]. An increase in secondary and tertiary branching with consequent enhancement in seed yield through increased number of infloresence and siliquae per plant was observed with the application of Mixatalol (a mixture of long aliphatic alcohols varying in chain length from C_{24} to C_{32}) to Brassica plants as foliar spray [56]. The percentage of immature siliquae and shattering of siliquae decreased with this treatment. Mixtalol increased total dry matter of plants, partitioning coefficient, and harvest index. The contents of starch, protein, and oil were also higher in seeds from mixtalol treated plants.

The maximum plant height (169.1 cm), number of primary branches per plant (8.2), seed yield (2031 kg/ha), stover yield (5752 kg/ha), harvest index (26.1%), oil content (42%), and net returns (Rs. 20,471/ ha) were recorded with thiourea (Shrama and Jain, 2003). At Bawal and Morena, highest seed yield (2060 kg/ha) was obtained with 40 kg S/ ha + thiourea (0.1%). At Sriganganagar, significantly higher seed yield (1883 kg/ha) was recorded on a par with 40 kg S/ha + thiourea (0.05%), urea (2%), H_2SO_4 (0.1%), and 40 kg S/ha. 40 kg S/ha + thiourea (0.1%) resulted into 17.67% higher seed yield over no spray. The highest oil content (35.9%) was recorded with thiourea 0.1% spray. Glucosinolate content ranged from 115 to 154 (µmole/g defatted meal) in different treatment (Table 12).

Table 12: Seed yield (kg/ha) and net returns (Rs./ha) of mustard as influenced by foliar application of agrochemicals at different locations

Treatment	S. K. Nagar		Sriganganagar	Ludhiana		
	Seed yield (kg/ha)	Net returns over control	Seed yield (kg/ha)	Oil content (%)	Oil yield (kg/ha)	Glucosinolate (µ mole/g defatted meal)
Control	1707	—	1604	34.7	375	130
Thiourea (0.1%)	2087	3226	1696	35.9	429	142
S @ 40 kg/ ha	2249	6712	1799	35.2	405	149
S @ 40 kg/ ha + Thiourea (0.1%)	2039	4070	1883	33.4	411	134
Urea (2%)	2019	5409	1845	34.7	396	124
ZnSO4 (0.5%)	1921	4622	1667	33.2	372	126

Boric acid (0.1%)	1928	3418	1650	34.3	387	115
CD at 5%	150	—	158	—	—	—

Source: AICRP-RM, 2003 [5].

IMPACT OF LOW MONETARY AGRO-TECHNIQUES ON MUSTARD PRO-DUCTIVITY

Agricultural inputs like fertilizer, irrigation, insecticides, pesticides, and herbicides, and so forth, are very expensive. Some nonmonetary or low monetary inputs can enhance the yield considerably with a slight increase in the cost of cultivation. There are a number of low monetary agro techniques which enhance the mustard yield considerably (Table 13). For harvesting the maximum yield of rapeseed-mustard at a given situation, all the production technologies, like, soil amendments, thinning, nutrient supply, sowing direction, irrigation, plant protection, and so forth should be planned well in advance. At Bharatpur, highest seed yield (1464 kg/ha) was recorded with the application of recommended practice (RP) + thinning at 15 and 25 DAS + detopping at bud-initiation stage followed by RP + thinning at 15 and 25 DAS.

Table 13: Effect of low monetary agrotechniques on seed yield and oil content of mustard at Bharatpur during 1997-1998

Treatments	Seed yield (kg/ha)	% increase over local practice	Oil content (%)	Oil yield (kg/ha)
Local Practice (T1)	1200	—	40.3	463
RP (No thinning and gypsum) (T2)	1371	14.2	40.3	525
RP + thinning at 15 & 25 DAS (T3)	1407	17.3	40.5	560
T3 + N-S sowing (T4)	1376	14.7	40.7	560
T3 + Removal of 4th row and 4th plant (T5)	1156	3.7	40.4	467

T5 + 56.75% N as top dressing (T6)	1073	10.6	40.3	432
T3 + I irrigation at 40–50 DAS (T7)	1232	2.7	40.6	500
T1 + 200 kg gypsum/ ha (T8)	1217	1.4	40.9	500
T3 + removal of 4 older leaves (T9)	1343	11.9	40.5	544
RP + de-topping at bud-initiation stage (T10)	1464	22	40.7	596

Source: AICRP-RM, 1998 [12].

FUTURE LINE OF RESEARCH

Rapeseed-mustard will continue to contribute considerably to the oilseed bowl of the country. A streamlined research programme for rapeseed-mustard should be focused on the below-mentioned points.

- Horizontal and vertical intensification in rapeseed-mustard production needs to be done for self-sufficiency in oilseed production. It is possible through varietal improvement and introduction of mustard in nontraditional areas.

- An optimum agronomic package of practices for high yielding and insect, pest, and disease resistant varieties, along with the upcoming hybrids needs to be worked out.

- Adoption of site-specific nutrient management (SSNM), precision agriculture, and conservation agriculture can bring more profits to the mustard growers.

- An integrated weed management approach needs to be developed for problematic and parasitic weeds in mustard. Orobanche is becoming a serious constraint and for its management a holistic approach which includes GM techniques needs to be explored.

- Suitable crop models and simulation for various inputs like water and nutrients will be helpful to target the most productive and most potential mustard growing zones of India.

CONCLUSIONS

The tremendous increase in oilseed production is attributed to the development of high yielding varieties coupled with improved production technology, their widespread adoption and good support price. To meet the ever-growing demand of oil in the country, the gap is to be bridged through management techniques. The vertical growth in mustard production can be brought by exploiting the available genetic resources with breeding and biotechnological tools which will break the yield barriers. Horizontal growth in rapeseed-mustard can be brought in those rapeseed-mustard growing areas/districts of the country, wherever, the yield is lower than the national average. Production technologies for different agroecological cropping systems, crop growing situations like intercropping, salinity, rainfall, and so forth, under unutilized farm situations like rice-fallows, mustard to be followed after cotton, sugarcane, soyabean, and so forth, and mustard as a paira crop in rice with lathyrus, lentil or any other competing rabi crop in traditional and nontraditional areas, need to be worked out. It is estimated that at least 1 million hectares can be brought under cultivation, through adoption of such cropping systems.

Proper land preparation, proper time of sowing, selection of better quality seeds, and so forth are always neglected. Fertilizer application is little or nonexistent leading to poor productivity. Whether little is spent on fertilizer input goes entirely on nitrogenous fertilizers. This results in a big gap between requirement and production of mustard in India. Therefore site-specific nutrient management through soil-test recommendation based should be adopted to improve upon the existing yield levels obtained at farmers field. Optimum crop geometry, balanced NPK fertilizers, intercultural operations, and inclusion of farmyard manure are the building blocks for achieving the utmost yield targets of rapeseed-mustard. Effective management of natural resources, integrated approach to plant-water, nutrient and pest management and extension of rapeseed-mustard cultivation to newer areas under different cropping systems will play a key role in further increasing and stabilizing the productivity and production of rapeseed-mustard to realize 24 million tonnes of oilseed by 2020 AD.

REFERENCES

1. India. Directorate of Economics and Statistics, Agricultural Statistics at a Glance, Department of Agricultural and cooperation. Ministary of Agriculture, Government of India, 2010.

2. J. S. Chauhan, K. H. Singh, and A. Kumar, "Compendium of Rapeseed-mustard varieties notified in India," Directorate of Rapeseed-Mustard Research, Bharatpur, Rajasthan, pp. 7–13, 2006.

3. M. A. Asoodari, A. R. Barzegar, and A. R. Eftekhar, "Effect of different tillage and rotation on crop performance," International Journal of Agrcultural Biology, vol. 3, no. 4, article 476, 2001.

4. A. L. Rathore, A. R. Pal, and K. K. Sahu, "Tillage and mulching effects on water use, root growth and yield of rainfed mustard and chickpea grown after lowland rice," Journal of Science of Food and Agriculture, vol. 78, no. 2, pp. 149–161, 1999.

5. AICRP-RM, Annual Progress Report of National Research Centre on Rapeseed-mustard.2002-2003, pp. 11–14, 2003.

6. T. Nagra, R. E. Phillip, and J. E. Legett, "Diffusion and mass flow of nitrate-N into corn roots under field conditions," Agronomy Journal, vol. 68, pp. 67–72, 1976.

7. R. L. Blevins, M. S. Smith, and G. W. Thomas, "Change in soil properties under no tillage," in No Tillage Agriculture, pp. 190–230, New York, NY, USA, 1984.

8. S. Snapp, R. Price, and M. Morton, "Seed priming of winter annual cover crops improves germination and emergence," Agronomy Journal, vol. 100, no. 5, pp. 1506–1510, 2008.

9. N. A. Khan and S.O. Aziz, "Response of mustard to seed treatment with pyridioxine and basal and foliar application of nitrogen and phosphorus," Journal of Plant Nutrition, vol. 16, no. 9, pp. 1651–1659, 1993.

10. A. Kumar, B. Singh, Yashpal, and J. S. Yadava, "Effect of sowing time and crop geometry on tetralocular Indian mustard," Indian Journal of Agricultural Sciences, vol. 62, no. 4, pp. 258–262, 2001.

11. AICRP-RM, Annual Progress Report of All India Coordinated Research Project on Rapeseed-Mustard, pp. 97–147, 1997.

12. AICRP-RM, Annual Progress Report of National Research Centre on Rapeseed-mustard. 1997-98, pp. 8–18, 1998.

13. R. Gupta, R. P. Avasthi, and S.J. Kolte, "Influence of sowing dates on the incidence of Sclerotinia stem rot of Rapeseed-mustard," Annals of Plant Protection Sciences, vol. 12, no. 1, pp. 223–224, 2004.

14. AICRP-RM, Annual Progress Report of All India Coordinated Research Project on Rapeseed-Mustard, pp. A1–16, 2007.

15. AICRP-RM, Annual Progress Report of All India Coordinated Research Project on Rapeseed-Mustard, pp. A1–44, 1999.

16. G. M. Khan and S. K. Agarwal, "Influence of sowing methods, moisture Stress and nitrogen levels on growth, yield components and seed yield of mustard," Indian Journal of Agricultural Science, vol. 55, no. 5, pp. 324–327, 1985.

17. M. J. Khan, R. A. Khattak, and M. A. Khan, "Influence of sowing methods on the production of canola grown in saline field," Pakistan Journal of Biological Sciences, vol. 3, no. 4, pp. 687–691, 2000.

18. F. C. Oad, B. K. Solangi, M. A. Samo, A. A. Lakho, Hassan-Ul-Zia, and N. L. Oad, "Growth, yield and relationship of Rapeseed (Brassica napus L.) under different row spacing," International Journal of Agriculture and Biology, vol. 3, no. 4, pp. 475–476, 2001.

19. H. P. Sierts and G. Geister, "Yield components stability in winter rape (Brassica napus L.) as a function of competition within the crop," in Proceedings of the 7th International Rapeseed Congress, p. 182, Poznan, Poland, May 1987.

20. AICRP-RM, Annual Progress Report of All India Coordinated Research Project on Rapeseed-Mustard, pp. 97–144, 1996.

21. AICRP-RM, Annual Progress Report of National Research Centre on Rapeseed-mustard. 2001-02, pp. 29–34, 2002.

22. K. K. Singh and K. S. Rathi, "Dry matter production and productivity as influenced by staggered sowing of mustard intercropped at different row ratios with chickpea," Journal of Agronomy and Crop Science, vol. 189, no. 3, pp. 169–175, 2003.

23. S. Chitale and M. C. Bhambri, "Response of Rapeseed-mustard to crop geometry, nutrient supply, farmyard manure and

interculture—a review," Ecology, Environment and Conservation, vol. 7, no. 4, pp. 387–396, 2001. View at Scopus

24. R. Sharma, K. S. Thakur, and P. Chopra, "Response of N and spacing on production of Ethopian mustard under mid-hill conditions of Himachal Pradesh," Research on Crops, vol. 8, no. 1, pp. 65–68, 2007.

25. D. Sah, J. S. Bohra, and D. N. Shukla, "Effect of N, P, S on growth attributes and nutrient uptake of mustard," Crop Research, vol. 31, no. 1, pp. 234–236, 2006.

26. M. L. Reager, S. K. Sharma, and R. S. Yadav, "Yield attributes, yield and nutrient uptake of Indian mustard (Brassica juncea) as influenced by N levels and its split application in arid Western Rajasthan,"Indian Journal of Agronomy, vol. 51, no. 3, pp. 213–216, 2006.

27. A. S. Sidhu and K. S. Sandhu, "Response of mustard to method of N application and timing of first irrigation," Journal of Indian Society of Soil Science, vol. 43, no. 3, pp. 331–334, 1995.

28. H. L. S. Tandon, S Research and Agricultural Production in India, Fertilizer Development and Consultation Organization, New Delhi, India, 2nd edition, 1986.

29. M. R. J. Holmes, Nutrition of the Oilseed Rape Crops, Applied science publishers, Essex, UK, 1980, In TSI/FAI/IFA Symposium.

30. M. S. Aulakh and N. S. Pasricha, "S fertilization of oilseeds for yield and quality," in Proceedings of the TSI-FAI symposium. S in agriculture-S-11/3, 1988.

31. J. C. Katyal, K. L. Sharma, and K. Srinivas, S in Indian agriculture. pp. KS-2/1-2/12, 1997.

32. AICRP-RM, Annual Progress Report of All India Coordinated Research Project on Rapeseed-Mustard, pp. A1–22, 2007.

33. AICRP-RM, Annual Progress Report of National Research Centre on Rapeseed-mustard. 1999-2000, pp. 24, 2000.

34. M. Gupta and R. D. Kaushik, "Effect of saline irrigation water and Zn on the concentration and uptake of Zn by mustard," in Proceedings of the 18th World Congress of Soil science, Philadelphia, Pa, USA, July 2006.

35. AICRP-RM, Annual Progress Report of All India Coordinated Research Project on Rapeseed-Mustard, pp. A1–28, 2005.

36. AICRP-RM, Annual Progress Report of National Research Centre on Rapeseed-mustard. 2004-05, pp. 9–11, 2005.

37. R. Prasad, S. N. Sharma, S. Singh, and R. Lakshaman, "Agronomic practices for increasing nutrient use efficiency and sustained crop production," in Proceedings of the National Seminar on Resource Management for Sustainable Production, New Delhi, India, February 1992.

38. M. A. Bhat, Singh, Room, and D. Dash, "Effect of INM on uptake and use efficiency of N and S in Indian mustard on an inceptisol," Crop Research, vol. 30, pp. 23–25, 2005.

39. B. S. Kumpawat, "Integrated nutrient management for maize-mustard cropping system," Indian Journal of Agronomy, vol. 49, pp. 4–7, 2004.

40. K. M. Hati, A. K. Mishra, K. G. Mandal, P. K. Ghosh, and K. K. Bandopadhyay, "Irrigation and nutrient management effect on soil physical properties under soybean-mustard cropping system," Agricultural Water Management, vol. 85, no. 3, pp. 279–286, 2006.

41. N. Pandey, R. S. Tripathi, and B. N. Mittra, "N, P and water management in greengram-rice-mustard cropping system," Annals of Agricultural Reseaarch, vol. 25, no. 2, pp. 298–302, 2004.

42. S. S. Parihar, "Influence of N and irrigation schedule on yield, water use and economics of summer rice,"International Journal of Tropical Agriculture, vol. 19, no. 1–4, pp. 157–162, 2001.

43. R. K. Ghosh, P. Bandopadhyay, and N. Mukhopadhyay, "Performance of Rapeseed-mustard cultivars under various moisture regimes on the Gangetic Alluvial Plain of West Bengal," Journal of Agronomy and Crop Sciences, vol. 173, no. 1, pp. 5–10, 1994.

44. S. K. Yadav, K. Chander, and D. P. Singh, "Response of late-sown mustard to irrigation and N," The Journal of Agricultural Sciences, vol. 123, pp. 219–224, 1994.

45. B. B. Panda, S. K. Bandyopadhyay, and Y. S. Shivay, "Effect of irrigation level, sowing dates and varieties on yield attributes, yield, consumptive water use and water-use efficiency of Indian mustard," Indian Journal of Agricultural Sciences, vol. 74, no. 6,

pp. 339–342, 2004.

46. I. Piri, "Effect of irrigation on yield, quality and water-use-efficiency of Indian mustard," in Proceedings of the 14th Australian Society of Agronomy Conference, Adelaide, Australia, September 2008.

47. Gangasaran and G. Giri, "Growth and yield of mustard as influenced by irrigation and plant population," Annals of Agricultural Research, vol. 7, no. 1, pp. 68–74, 1986.

48. C. P. S. Chauhan and R. B. Singh, "Mustard performs well even with saline irrigation," Indian Farming, vol. 42, pp. 17–20, 2004.

49. M. A. Kahlown, M. Akram, Z. A. Soomro, and W. D. Kemper, "Prospectus of growing barley and mustard with saline ground water irrigation in fine and coarse textured soils of Cholistan desert,"Irrigation and Drainage, vol. 51, no. 4, pp. 328–338, 2008.

50. D. K. Majumdar, "Effect of supplementary saline irrigation and applied N on the performance of dryland seeded Indian mustard," Experimental Agriculture, vol. 31, pp. 423–428, 1995.

51. J. Pandey and B. N. Mishra, "Effect of weed management practices in a rice-mustard-mungbean cropping system on weeds and yield of crops," Annals of Agricultural Reaseach, vol. 24, no. 4, pp. 36–39, 2003.

52. B. R. Bazaya, D. Kachroo, and R. K. Jat, "Integrated weed management in mustard (Brassica juncea),"Indian Jounal of Weed Science, vol. 38, no. 1-2, pp. 16–19, 2006.

53. S. Kumar, "Identification of trap crop for reducing broomrape infestation in the succeeding mustard,"Agronomy Digest, vol. 2, pp. 99–101, 2002.

54. M. Mobin, H. R. Ansari, and N. A. Khan, "Timing of GA3 application to indian mustard: DM distribution, growth analysis and nutrient uptake," Journal of Agronomy, vol. 6, no. 1, pp. 53–60, 2007.

55. N. A. Khan, H. R. Ansari, and Samiullah, "Effect of gibberellic acid spray during ontogeny of mustard on growth, nutrient uptake and yield characteristics," Journal of Agronomy and Crop Science, vol. 181, no. 1, pp. 61–63, 1998.

56. R. C. Setia, Richa, N. Setia, K. L. Ahuja, and C. P. Malik, "Effect of Mixtalol on growth, yield and yield components of Indian

mustard (Brassica juncea)," Plant Growth Regulation, vol. 8, no. 2, pp. 185–192, 1989.

Citations

CHAPTER 1

Subhash Babu, D.S.Rana, G.S.Yadav, Raghavendra Singh, and S.K.Yadav, A Review on Recycling of Sunflower Residue for Sustaining Soil Health, doi.org/10.1155/2014/601049.

CHAPTER 2

R. Saha, R. S. Chaudhary, and J. Somasundaram, "Soil Health Management under Hill Agroecosystem of North East India," Applied and Environmental Soil Science, vol. 2012, Article ID 696174, 9 pages, 2012. doi:10.1155/2012/696174.

CHAPTER 3

G. B. Huang, Z. Z. Luo, L. L. Li, et al., "Effects of Stubble Management on Soil Fertility and Crop Yield of Rainfed Area in Western Loess Plateau, China," Applied and Environmental Soil Science, vol. 2012, Article ID 256312, 9 pages, 2012. doi:10.1155/2012/256312.

CHAPTER 4

Mark A. Liebig, David W. Archer, and Don L. Tanaka, "Crop Diversity Effects on Near-Surface Soil Condition under Dryland Agriculture," Applied and Environmental Soil Science, vol. 2014, Article ID 703460, 7 pages, 2014. doi:10.1155/2014/703460.

CHAPTER 5

Wolde Mekuria and Andrew Noble, "The Role of Biochar in Ameliorating Disturbed Soils and Sequestering Soil Carbon in Tropical Agricultural Production Systems," Applied and Environmental Soil Science, vol. 2013, Article ID 354965, 10 pages, 2013. doi:10.1155/2013/354965.

CHAPTER 6

Motior M. Rahman, Aminul M. Islam, Sofian M. Azirun, and Amru N. Boyce, "Tropical Legume Crop Rotation and Nitrogen Fertilizer Effects on Agronomic and Nitrogen Efficiency of Rice," The Scientific World Journal, vol. 2014, Article ID 490841, 11 pages, 2014. doi:10.1155/2014/490841.

CHAPTER 7

R. Moussadek, R. Mrabet, R. Dahan, A. Zouahri, M. El Mourid, and E. Van Ranst, "Tillage System Affects Soil Organic Carbon Storage and Quality in Central Morocco," Applied and Environmental Soil Science, vol. 2014, Article ID 654796, 8 pages, 2014. doi:10.1155/2014/654796.

CHAPTER 8

Gebreyesus Brhane Tesfahunegn, Soil Quality Assessment Strategies for Evaluating Soil Degradation in Northern Ethiopia, http://dx.doi.org/10.1155/2014/646502.

CHAPTER 9

Md. Khairul Alam, Md. Monirul Islam, Nazmus Salahin, and Mirza Hasanuzzaman, "Effect of Tillage Practices on Soil Properties and Crop Productivity in Wheat-Mungbean-Rice Cropping System under Subtropical Climatic Conditions," The Scientific World Journal, vol. 2014, Article ID 437283, 15 pages, 2014. doi:10.1155/2014/437283.

CHAPTER 10

Kapila Shekhawat, S. S. Rathore, O. P. Premi, B. K. Kandpal, and J. S. Chauhan, "Advances in Agronomic Management of Indian Mustard (Brassica juncea (L.) Czernj. Cosson): An Overview," International Journal of Agronomy, vol. 2012, Article ID 408284, 14 pages, 2012. doi:10.1155/2012/408284.

Index